Microcomputer Communications

Martin Gandy

PUBLISHED BY NCC PUBLICATIONS

British Library Cataloguing in Publication Data

Gandy, Martin
  Microcomputer communications
  1. Telecommunication systems – Data processing
  2. Microcomputers
  I. Title
  621.38'0413   TK51055
  ISBN 0-85012-495-6

© THE NATIONAL COMPUTING CENTRE LIMITED, 1985

All rights reserved. No part of this publication may be reproduced, stored in a retrieval system, or transmitted, in any form or by any means, without the prior permission of The National Computing Centre.

First published in 1985 by:

NCC Publications, The National Computing Centre Limited, Oxford Road, Manchester M1 7ED, England.

Typeset in 11pt Times Roman by UPS Blackburn Limited, 76-80 Northgate, Blackburn, Lancashire, and printed by Hobbs the Printers of Southampton.

ISBN 0-85012-495-6

Introduction

Rapid advances in microcomputer design and a growing awareness of its potential have opened up new application areas. Word processing, spreadsheet and database packages are now well known, and a type of application previously restricted to mainframe systems – communications – has caught the imagination of the microcomputer user. In this book the opportunities open to the microcomputer user through communications links are outlined, with practical guidelines on how these links can be installed.

Today's 16-bit hardware and operating systems offer considerably improved performance over earlier designs and provide the basis for developing the microcomputer into a sophisticated communications device. This book follows the development of the microcomputer and reviews the desirable features required to give the microcomputer useful communications capabilities.

The book first looks at data formats, and the transmission of data at the physical and electrical level. The broad categories of communication are examined next in order of increasing sophistication. These are the microcomputer as a terminal emulator, file transfer and integrated microcomputer-to-host computer communications software. The later sections of the book deal with the microcomputer in local area networks and wide area networks. This includes specially designed high-speed bus and ring networks for a single site, and conventional communications networks based on modems and analogue telephone lines.

Microcomputers are found in many different environments and the users in turn have quite different communications needs. For example, in an office system, the main requirement is for transferring documents between machines. In production management however, access to databases which may be distributed around the site will be the prime need. Scientific and computer support staff are primarily terminal users and are interested in multi-terminal emulations which can access a variety of host computers. There are also some communications facilities which are attractive to all types of user. The most familiar example is electronic mail which can solve the problems of contacting staff by telephone or by the internal post.

The main barrier to successful interworking has been the lack of standards. Fortunately progress is now being made on international communications standards through the International Standards Organisation (ISO). The structure of the ISO model is outlined and the present position of standards work is described.

A great variety of equipment and techniques can be encountered in a computer network and this can lead to an equally wide range of problems. A list of these pitfalls has been compiled from users' experiences and some guidelines are given which should help to avoid them.

Clearly microcomputer communications is a key subject for both the end user and dp professional. The need is growing and once the number of microcomputers in an establishment exceeds a 'critical mass' the benefits can be very significant. The techniques are still at an early stage in their development, but there are some useful products available now, and these are being enhanced to give more flexibility and make them easier to use.

Acknowledgements

I wish to thank the following organisations for giving up their time and allowing me to benefit from their practical experience:

>Bristol University Computer Centre
>Edinburgh University Regional Computer Centre
>University of Manchester Regional Computer Centre

The following organisations are thanked for providing operating manuals and technical data, and in some cases for supplying equipment for evaluation:

>British Telecom
>Camtec Electronics Ltd
>Gandalf Digital Communications Ltd
>Informatics General (UK) Ltd
>International Computers Ltd
>Racal-Milgo Ltd
>Ultracomp Ltd
>Vuman Ltd

I would also like to thank the following individuals for their help, advice and comments on the text:

>P R D Scott
>K C E Gee
>P Hardy
>A Gardner

Finally the Centre acknowledges with thanks the support provided by the Electronics and Avionics Board of the Department of Trade and Industry.

Contents

	Page
Introduction	
Acknowledgements	
1 The Role of the Microcomputer	13
Introduction	13
Hardware	16
Software	20
Communications Issues	24
2 Data Transmission – The Physical Link	27
Introduction to Data Formats	27
Asynchronous Transmission	28
Synchronous Transmission	30
Serial Interface Standards	32
Making the Connection	36
Summary	43
3 The Serial Interface in Direct Connections	45
Cross Connections	45
How Far, How Fast?	48
New Electrical Standards	51
The OSI Seven-Layer Model	54

4 Terminal Emulation 59

Why Terminal Emulation? 59
Choosing a Terminal Emulation 60
Some Terminal Emulation Products 62

5 File Transfer 67

Introduction 67
File Transfer – The Problem of Incompatibility 67
File Transfer – Transmission Problems 70
File Transfer Products – A Classification 72
File Transfer Using System Utilities 75
File Transfer Products 77
Case Studies 81
File Transfer Standards 86
Protocol Converters 96

6 Local Area Networks 97

The Need for Local Area Networks 97
Local Area Networks 99

7 Wide Area Networks 129

Introduction 129
Proprietary Network Architectures 131
Public Data Switching Networks 132
The Packet Assembler/Disassembler 133
The Data Switch 140
Private Networks – Conclusions 141
Connection to Public Data Services 142

8 Problems in Communications 145

Some of the Pitfalls 145
Synchronisation during Interactive Processing 145
Flow Control and Character Loss 146
Non-Transparency of Circuit 147
Deficiencies in Screen Format or Resolution 148
Character Set Problems 148
Keyboard Limitations 149
Incompatible Control Codes 150

		Poor Response Times and Excessive Delay in Keyboard Echo	151

9 Conclusions 153

Appendix

    1 References and Bibliography   155
    2 Suppliers of Communications Software
         and Equipment   157
    3 Abbreviations   165
    4 Glossary of Communications Terms   169

Index 183

1 The Role of the Microcomputer

INTRODUCTION

There are clear phases in the history of computing. The 1950s and 1960s were dominated by central mainframes running data processing jobs in batch mode. The end user of the data had very little contact with the computer, which was managed and run in a separate department by professional staff including systems analysts, programmers, operators and clerical personnel. The 1970s saw the emergence of the minicomputer, which placed a strong emphasis on interactive programming using visual display terminals (VDTs). New areas for computing in the small business sector opened up, where costs of changing to a computer system had previously been too high. Mainframe computers still held the dominant position in terms of value of sales however, and probably will for some time. Another major change had become clear by the 1980s: the microcomputer had burst onto the scene, first as a domestic or laboratory tool, and then as a serious business machine.

The changes have only been possible because of the rapid developments in technology, particularly in large-scale integration and mass storage. The development of smaller and more powerful computers has not taken much work away from the central computer services department; instead new application areas, not possible with the early mainframe computers, have been opened up. It is likely that some applications – for example in banking, insurance and management – will remain dependent on mainframes, centralised, and under the responsibility of professional data processing staff.

A good example of the progress being made in large-scale integration is the short history of random access memory (RAM) integrated circuits ('chips'). For example, progress has been made by doubling the dimensions of the silicon chip to give four times the area and four times the component count. So far we have seen the state-of-the-art RAM increase in memory capacity in steps from 1K (1970), 4K (1974), 16K (1976), 64K (1980) and to a 256K chip during 1984. The 1 Mbyte chip is expected by 1986.

This rapid development has been made possible by increasing the purity of the silicon crystal growing process, so that larger fault-free areas can be obtained. Improved mask production using electron beam techniques has also increased the packing density that can be obtained. The scaling-up process of circuit layout has been a relatively simple problem. This continuing evolution of large-scale integration technology has led to:

— greater memory capacity per chip;

— more powerful CPUs;

— smaller-size printed circuit boards with fewer components;

— lower power consumption;

— greater reliability;

— faster performance;

— more powerful operating systems;

— multi-tasking operating systems;

— better applications.

If the aim of information processing can be stated as getting the right information to the right person at the right time in the right place in the right form and at the right cost, then the microcomputer is helping to meet this aim. The first microcomputers were stand-alone devices with limited power and memory, but the potential of the microcomputer as a business machine was soon evident: and the number of effective business application programs rapidly increased. This caught the imagination of the end user of information, and the microcomputer was quickly accepted as an office machine and management tool.

THE ROLE OF THE MICROCOMPUTER 15

The key features in the microcomputer that have made this possible are:

— low cost;

— cheap mass storage;

— integrated screen and CPU;

— small number of CPU chips;

— small number of operating systems;

— high quality of software, which can run on many different machines.

The cost of a typical business microcomputer is now comparable with that of a computer terminal ten years ago, when discrete components were still widely used. Cheap mass storage is another technological development which, although not so dramatic as chip technology, is still impressive. Personal mass storage using Winchester technology is now affordable, in units of 5, 10, 20 and 100 Mbytes, with 500 Mbytes promised in 1985. These storage volumes were not available to the mainframe market 20 years ago.

The integrated screen and CPU connected by the internal high-speed bus is another key feature which has made possible some of the new applications, such as word processing and spreadsheets. Traditional data processing was adequately served by terminals which could be updated from the central host computer over a serial line connection. The fastest transmission speed usually possible over local connections is 9600 baud, which takes two seconds to fill a screen. This is too slow for spreadsheets and graphics packages, and poor for word processing. The microcomputer, however, can update a screen an order of magnitude faster than this because the display system and CPU are connected by a high-speed bus.

The last three factors are closely linked together. As integrated circuit development is very costly, relatively few microprocessor chips, the heart of the microcomputer, have been developed and used successfully. Hence only a few operating systems able to run on a variety of different machines have become widely established.

With many common families of machines coming onto the market the incentive to write software was very great and the competition fierce. This has resulted in some excellent software products available at very lost cost. In contrast, in the mainframe market every machine is different, and there is little possibility of transferring software from one machine to another. This has kept software costs high, and the availability of good software low.

As the business community accepted the microcomputer as an essential business machine, the limitations of a stand-alone machine were soon realised. Information is of little use until it is distributed, and microcomputers, linked together and linked to central computing facilities, can improve communications flow throughout a business.

Direct links between computing equipment have been in widespread use since the earliest days of computing development. These links either used special data-quality cable for on-site applications, or used the telephone network for longer-distance communications. This type of computer network does not provide an easy or cheap way to pass information between users. A new development, the local area network, was introduced to meet this new need and is the key to office automation. The main feature of a local area network is a data highway which is common to all users. Network software allows the users to access shared resources, such as high-cost printers, and can offer important new services such as electronic mail.

HARDWARE

All microcomputers (and other digital computers) have the same basic structure, as shown in Figure 1.1. Information is represented by binary digits (BITS), which can only take the values 0 or 1. This coding scheme is used as it is very easy to represent logical 0 accurately as approximately 0 volts and logical 1 as approximately 5 volts. These are the most popular voltage levels for digital electronic components. By choosing a threshold midway between these voltage levels, a digital system will have very good immunity to interfering noise voltages. For example, levels of 1 volt and 4 volts will be clearly interpreted as logic levels 0 and 1 respectively.

To represent information adequately, binary digits must be

THE ROLE OF THE MICROCOMPUTER 17

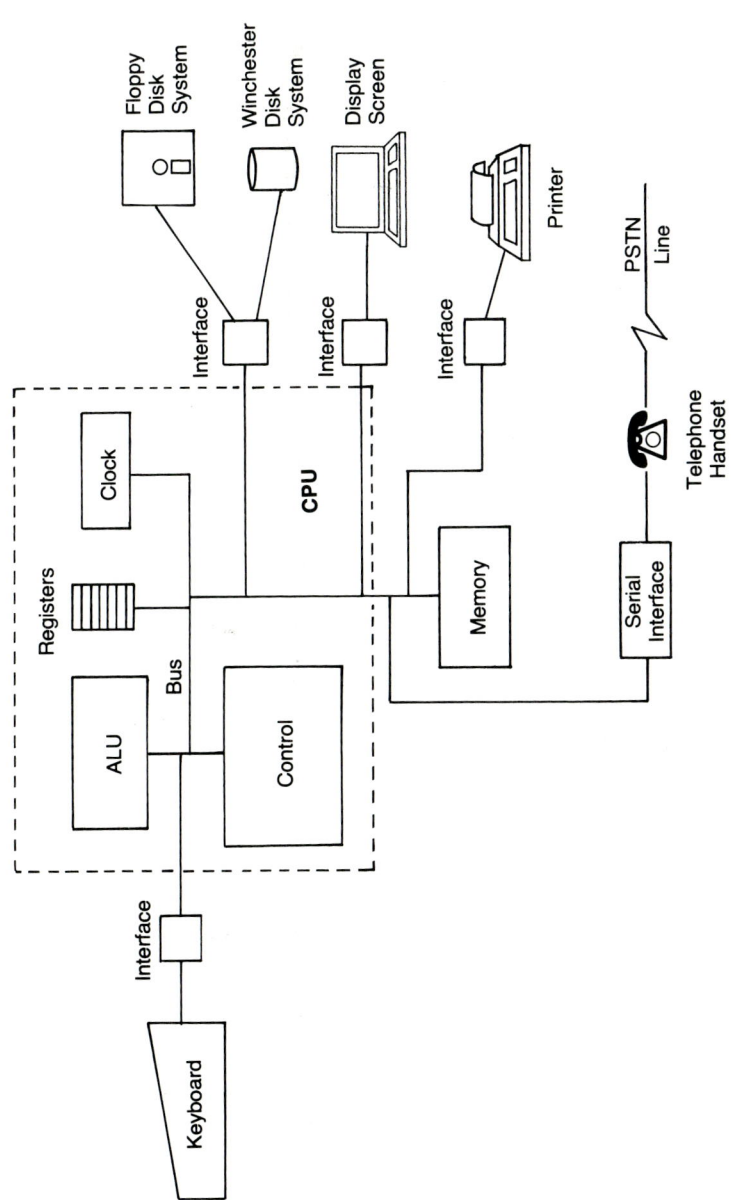

Figure 1.1 Microcomputer Architecture

grouped together into units termed words. The word length is one of the factors which identifies the size and power of a computer, with word lengths of 8 or 16 bits typically found in microcomputers, and word lengths of 32 or more bits found in mainframes. This difference is becoming less clear as the rapid progress in large-scale integration is pushing the single chip word length closer to the mainframe, with 32 bits now possible. The other main difference between the microcomputer and the mainframe is the use of faster circuitry in the larger machines. Faster circuitry uses more power, which in turn needs more space and weight for cooling and power supplies.

The main components shown in Figure 1.1 are connected together by a set of lines termed a bus. This is responsible for carrying all control and data signals around the microcomputer and to the outside world. At the heart of the microcomputer is the central processing unit (CPU). This has a number of major components including the control unit, the arithmetic and logic unit (ALU), a set of internal storage units (registers) and a clock system. The ALU provides a number of primitive functions such as add, subtract, compare, shift and rotate. Each operation is uniquely identified by a binary code called an instruction. The microcomputer has a large number of different instructions called the instruction set. A more powerful computer will have a larger and more powerful instruction set than a low-cost, 8-bit microcomputer.

The CPU performs tasks by following a sequence of instructions and acting on data values. These are stored in the memory unit, which is connected to the CPU by the bus. There are two types of memory shown in Figure 1.1: read only memory (ROM) and random access memory (RAM). Each type of memory is an array of memory cells, each capable of storing one bit of information, normally arranged as 8-bit groups (bytes). Each byte has a unique address, which the CPU can access by issuing an address over the bus and then reading the bus data lines. In addition to address and data lines, the bus also has a number of control lines, and clock lines, to control data flows between CPU and memory. The difference between ROM and RAM memory elements is that the ROM contains permanent data whereas the RAM can be written to and read from.

THE ROLE OF THE MICROCOMPUTER

The number of memory elements that can be addressed depends on the word length and memory bus size. Most 8-bit machines have 16 address lines and so can address 64K bytes of data. More powerful 16-bit machines have more address lines, and may be able to address 16 Mbytes or more.

The computer operates on a fetch-decode-execute cycle, taking instructions in sequence from the memory. The CPU finds the address of the first instruction and then fetches this instruction from memory. After decoding the instruction, the ALU is given the control signals required to execute this particular instruction. Once completed, the CPU fetches the next instruction, which is decoded and executed. This process continues until a test tells the CPU that no more instructions are to be accessed.

To complete the microcomputer design, connections to the outside world must be provided. Several different external devices or peripherals are shown in Figure 1.1. These are connected to input/output (I/O) interfaces which are in turn connected to the CPU via the computer bus. A variety of different interfaces is needed to match the different characteristics of the peripheral devices to the bus. Some of the important I/O devices are the keyboard and display screen (which are usually an integral part of the microcomputer), floppy disk units, Winchester disks and printers. The final interface shown in Figure 1.1 is the serial line interface connected to the modem, which provides communications facilities. The operation of this interface and other communications interfaces forms a major part of this book.

There are only a relatively few CPU chips in use because of the high development costs. The first CPU chips were the 4-bit 4004 and 4040 units from Intel introduced in 1970. Although they have their uses these devices were not powerful enough for microcomputer design. These four-bit devices were followed up by the 8-bit 8008 in 1971 and an improved and very successful 8080 in 1973. The success of these devices led to several imitators producing similar 8-bit designs, including Zilog's Z80, Motorola's 6800 and MOS Technology's 6502. The Z80, which has the 8080 instruction set as a subset of its own more extensive instruction set, and the 6502 have been particularly successful in the number of microcomputers that have been designed around them. It is unlikely that there will be any significant development of these 8-bit CPUs as the development effort has now moved to 16- and 32-bit designs. The most important of the 16-bit CPUs are the 8088 and 8086 from Intel and the 6800 from Motorola. These are used for example in the IBM PC (8088) and the Apple Lisa (6800).

The earliest microcomputers had very limited performance and were strictly single-user, single-task machines. As microprocessor chips and associated interfacing and memory chips have become more powerful, a whole family of microcomputers has become available. This family includes scientific microcomputers, multi-emulation intelligent terminals, business systems, portable microcomputers, workstations, and multi-user, multi-tasking machines. The last two categories are most interesting in the context of this book because of their potential as communications devices. The multi-tasking machine has a major advantage over the single-job machine, as it can handle the communications interface whilst the user is running another application. This means, for example, that a telex message could be received while the user is running a spreadsheet program.

SOFTWARE

Operating Systems and Utilities

The microcomputer hardware cannot perform any useful tasks until programs (software) have been loaded into memory. Even the simplest of tasks, such as reading a character from the keyboard, requires an elementary program, and a sophisticated program like a spreadsheet is built up from a number of layers of more elementary programs. Most computers have three levels of software: the operating system, tools and utilities, and applications programs. Each level uses the services of the lower levels, breaking down complex tasks into simpler tasks. The operating system is the most basic software which interfaces directly to the hardware. The main jobs of the operating system are to look after the keyboard, the display screen, the disk system and the management of the memory. A number of useful subroutines are built into the operating system and these can be used by the higher-level software.

The second level of software includes a wide range of programs

THE ROLE OF THE MICROCOMPUTER 21

to help the user manage the system, develop new applications and run existing ones. One of the most important utility programs is the Copy facility which enables the user to manipulate disk files in a variety of ways. Other examples include the Text Editor for creating source programs or documents, and the Library facility for creating a library of commonly used subprograms. The other major class of software which operates at this intermediate level includes the high-level languages. A high-level language allows programs to be written in statements resembling English text; these are then converted to sequences of binary instructions taken from the instruction set of the target microcomputer chip. The most familiar language to microcomputer users is BASIC, which is the first-choice language because it is easy to learn and debug. More powerful languages were soon introduced for the common operating systems, including COBOL, PASCAL, FORTRAN, C, APL, FORTH, LISP and RPG 11.

At the highest software level are the application programs. These are the programs that allow useful tasks to be performed such as word processing, spreadsheets or data base management. These programs may have been written in a high-level language, and usually make use of the services provided by the operating system.

None of the popular operating systems – CP/M, MS-DOS, Unix and PC-DOS – support communications well. With the exception of Unix, these operating systems were designed as single-user, single-tasking operating systems and this is not the right environment for communications support. Unix and its derivatives (Zenix for example) have the right structure to support communications directly as there is a mechanism for installing device drivers which can be controlled concurrently with other tasks. Concurrent operating systems hold the key to better communications support, and one of the newer operating systems, concurrent CP/M, looks quite promising. Multi-tasking is only really successful with the more powerful 16- and 32-bit microcomputers. An earlier 8-bit solution, MP/M, has not been widely used. When microcomputers are linked together to provide a communication service to a group of users there are many more control requirements than are needed in stand-alone systems. These include:

— password access;

- file locking;
- file and text transfer;
- remote control.

The extra control requirements put a number of extra demands on the operating system, including:

- priority interrupt structure;
- multi-tasking and timing control;
- error detection and correction during file transfers;
- additional buffer space and control;
- alternate routeing.

Applications

From the user's point of view it is the quality and quantity of software that really counts. Whilst the keyboard layout and other hardware factors are important it is the software that sells microcomputers. The microcomputer is used in a variety of ways, which fall into the following broad categories:

- toy;
- controller;
- terminals;
- tool;
- computer.

Most of the business applications fall into the terminal and tool categories, with the controller function used in some industrial and medical environments, and the computer function used for teaching and research applications. Hopefully the microcomputer used as a toy is reserved for the home. The most important market at the moment is in office automation, which includes a very wide range of services:

- shared disk access;
- electronic mail;

- shared printer access;
- word processing and other business tools;
- access to external services via gateways;
- telephone support;
- voice recognition.

The office automation system demands a microcomputer that has more facilities than the personal computer, and may have an integrated feature telephone. Because of the range of services offered, this type of system has been termed a workstation. The external services which may be required include Telex, Teletex, Videotex (eg Prestel) and Mailbox systems (eg Telecom Gold).

The business user has some different requirements to the office automation user including financial modelling (spreadsheet), data base, graphics support and communication links to central computer services. Business software for the stand-alone microcomputer has been very highly developed and the spreadsheet and word processing packages have opened up new areas for the computer. It is the success of these software packages that has created the need for communication links between microcomputers and for connections to the mainframe. With these links in place it will be possible to transfer messages and data around the user group, and to access the central data base. The last point is an important development in the use of microcomputers. With good communications software the microcomputer user will be able to access corporate files and transfer selected fields for local processing. This should be done without needing to master the host computer operating system or file structure. Ideally the transfer should be set up by menu selections from the microcomputer screen, and completed whilst the machine is used for other purposes.

As distributed computing is extended to individual managers and clerical staff the problems of security and privacy become more serious. The distributed computing system must protect against unauthorised access to microcomputer software and mainframe data. Data integrity must be maintained which can be a major problem if a large number of users are allowed access to update mainframe data. The problem of disaster recovery is more difficult if files are distributed about the system and local transactions are not recorded centrally. There are other related security issues such as possible theft of microcomputers with their data and infringement of copyright agreements as software is passed around a network.

COMMUNICATIONS ISSUES

In this section the techniques described in the remainder of the book are introduced. As a starting point, consider the basic link between two machines, as shown in Figure 1.2. The diagram shows the terminal connected to the physical line by an interface. Wherever there is a connection between two systems via an interface, a set of rules is required to handle the data flow across the interface. A second end-to-end set of control rules is also required to manage the transfer of data at the application level, which is a higher-level requirement. The higher-level rules make use of the lower-level rules, forming a hierarchy. These rules or protocols can be broken down into more than just two levels, and this subject is returned to in the discussion on internal standards for communications, in particular the International Standards Organisation (ISO) 7-layer model for Open Systems Interconnection (OSI).

The physical line can be provided in many ways and most development work is now concentrated on networks of interlinked microcomputers and host computers, rather than the point-to-point link. The main techniques in use to provide a connection service are:

— direct-wire links on site;

— dial-up links via the PSTN;

— leased-line links;

— public switched networks;

— value-added networks;

— micronets;

— local area networks;

— broadband networks.

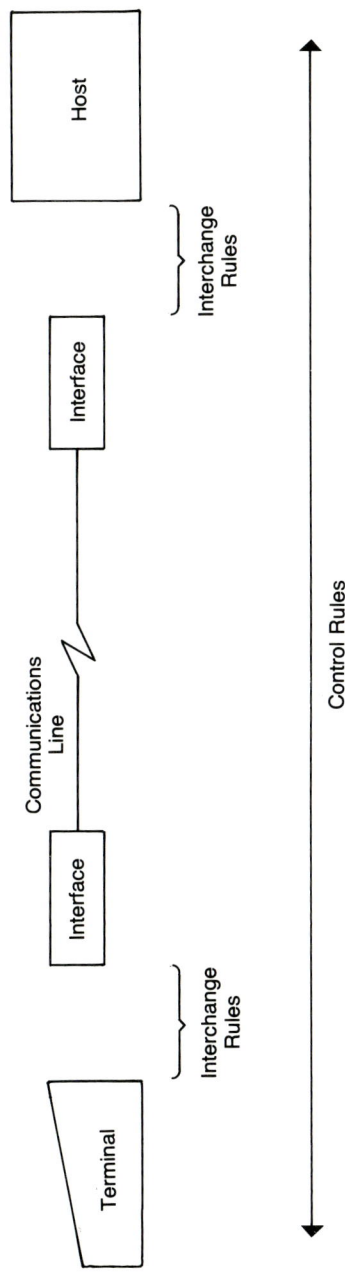

Figure 1.2 Symbolic Communications Link

These techniques are listed approximately in order of speed and complexity. The faster techniques in the local area network family are strictly limited in network size and would normally be installed on a single site, which could be an office or industrial complex.

The main problem in any network is in achieving compatibility. This means that any two machines from different suppliers should be able to interwork at all levels. This is a formidable task because every supplier favours different techniques at every conceivable level. The ISO standards work represents a major initiative on the part of the computer industry to help solve this problem of interworking.

2 Data Transmission – The Physical Link

INTRODUCTION TO DATA FORMATS

The transfer of information over considerable distances has a long history. Signalling using a binary system can be traced back to the 1830s when the Morse Code was invented by Samuel Morse for long-distance communications. By giving each letter a unique code consisting of a group of long and short tones, messages could be transmitted over great distances using telegraph lines or radio links. Today's computer communications use different character sets and codes, and developments in technology have led to higher transmission speeds and increasing sophistication in the communications facilities available.

Information is represented in digital computers by groups of binary digits or bits. As each character is represented by a number of bits, two methods of transmitting data between computer equipment have evolved: parallel and serial transmission. In parallel transmission all the bits representing a character are transmitted together over a group of wires, whereas in serial transmission each bit is transmitted in sequence starting with the most significant bit.

Parallel transmission of data is usually restricted to the internal operation of computers and between the main processor (CPU) and storage devices or other CPUs over short distances. Parallel transmission is also found in some computer-to-printer connections, and between computers and laboratory instruments. For all other communications applications, serial transmission is the norm.

The Morse Code was developed for transmission and reception

by human beings, and does the job very efficiently. The variable code length and limited number of characters represented make it unsuitable for computer data representation, and other character sets have been developed for this purpose. However, standardisation continues to be a problem, as will be seen later.

The interpretation of Morse signals requires a skilled operator: the individual character sequences must be recognised from the data stream by the difference in inter-bit period compared to the inter-character period. In computer communications two methods of synchronising the transmitter and receiver are used: asynchronous and synchronous transmission.

ASYNCHRONOUS TRANSMISSION

Asynchronous transmission was developed first for use with electromechanical teletypewriter devices. In this technique, data is transmitted character-by-character, with an undefined period between characters.

Each character must be recognised and the transmission clock derived as no timing information is carried by the line. This is achieved by putting the character into a standard frame as shown in Figure 2.1. The line remains idle between characters, which is the high or 1 state. The string of data bits which comprise the character is preceded by a start bit, one unit long, which is always low or 0 and followed by a stop bit, which may be 1, 1.5 or 2 units long and is always at the high or 1 level. The number of data bits may be 5 to 8, with an additional bit which is used to indicate parity.

Synchronisation depends on the receiver recognising the transition from idle to 0 as the start bit is received and then setting the bit sampling to occur at the middle of each data bit. To achieve high transmission speeds, modem serial interfaces have a crystal-controlled clock running at high speed at each end of the line. A number of preset speeds are available and these are configured to be the same at each end.

Serial interfaces must be configured so that each end of the link has the same format and bit rate. A mismatch will result in garbled characters, lost characters or a cross-hatched character (indicating parity error). The data format and bit rate is usually set by dual-in-

DATA TRANSMISSION – THE PHYSICAL LINK

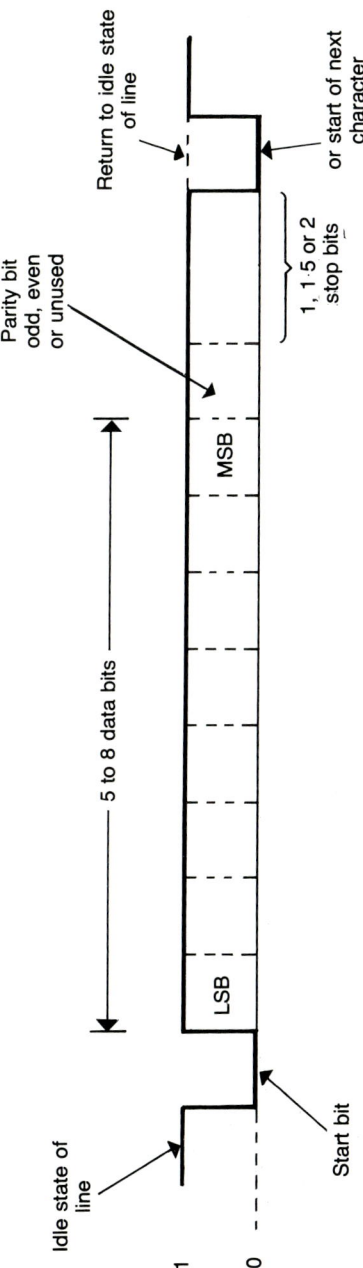

Figure 2.1 Asynchronous Character Format

line (dil) switches, but newer equipment allows the user to set up a default format which is available as the equipment is switched on with the facility to reconfigure the format under software control.

SYNCHRONOUS TRANSMISSION

Synchronous communication is characterised by the communications link carrying timing information in addition to data, and is the method of choice for higher speed transmission over the telephone network. It is also used almost exclusively by the mainframe suppliers for mainframe to terminal links.

Characters in a synchronous transmission are not separated by start and stop bits, but are joined together as a continuous bit stream. This requires a different form of synchronisation, which is based on a synchronising sequence being sent before data transmission can take place. A special bit sequence is sent and repeated by the transmitting end and the receiver scans the bit stream for the expected sequence which identifies the character frame. By repeating the sequence, protection against errors is provided. Once the receiver has locked onto the character frame, subsequent characters can be recognised. A typical synchronising sequence is shown in Figure 2.2.

In some ways a synchronous link is simpler to set up than is an asynchronous link. The terminal derives the timing from the line so there is no baud rate to set up, and the interface is not concerned with stop, start and parity bits. However, the stream of characters is organised into groups which are exchanged between machines using a set of rules known as a data link protocol. Each mainframe supplier has one or more data link protocols in use for their proprietary networks, and these are the subject of constant revision. It is this factor that makes interworking so often a difficult problem which will be examined in more detail later.

Having established the correct data format for a successful connection, the physical link must be made. At the physical link level, one type of electrical interface dominates the industry. This interface, known as RS-232 C or V.24, is a group of standards defining an interface between Data Terminal Equipment (DTE) and Data Communication Equipment (DCE). These terms will be referred to as Terminal (DTE) and Modem (DCE) where appropriate.

DATA TRANSMISSION – THE PHYSICAL LINK 31

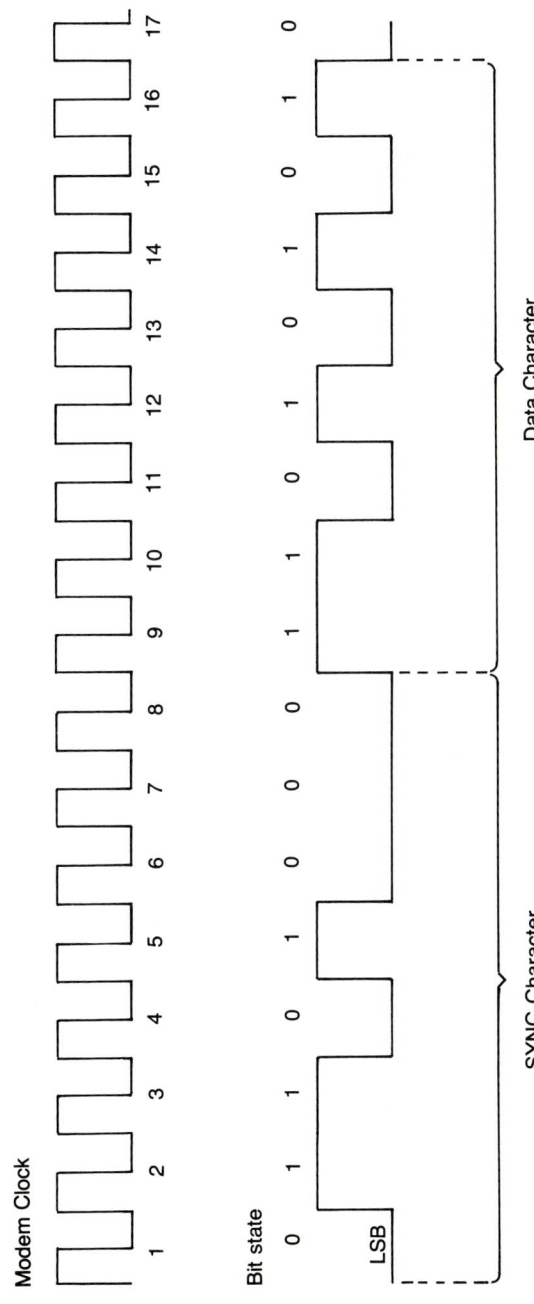

Figure 2.2 Synchronous Data Character Format

SERIAL INTERFACE STANDARDS

Linking different equipment together, either locally or across national boundaries, depends on the acceptance of standards. These standards have emerged from various standards bodies from time to time, and often with a considerable degree of overlap. The process is continuing as new technology leads to higher performance interfaces which must then be incorporated into new standards.

The first standard of interest, known in its latest revision as RS-232-C, was introduced by the Electronic Industries Association (EIA) in co-operation with the Bell System, modem manufacturers and computer manufacturers.

The serial interface is now commonly referred to as an RS-232 port. The standard was then adopted, with minor variations, by an international standards body, the CCITT, and appeared as two recommendations, V.24 and V.28. The interface, which was designed as a relatively low-speed link between terminals and modems, is now commonly used to connect terminals directly to microcomputers, to minicomputers and to computer networks.

The extended use of the interface standards has caused some problems of interpretation. As the signals are directional, a link between two microcomputers must be set up to look like a terminal connected to a modem. Further difficulties are caused by the freedom given to implementors in the use of the interface circuits. An interface usually only provides a subset of the defined circuits, and the circuits may be actively supported, passively supported (constant voltage provided) or unsupported (disconnected).

The serial interface will be described by reference to the international standards, and any significant differences between overlapping standards will be pointed out. The full titles and scopes of these documents are:

— CCITT Recommendation V.24 (1964, amended 1968, 1972, 1976): List of definitions of interchange circuits between data terminal equipment and data circuit terminating equipment.

  The document is short and concise, a factor which has led to difficulties in interpretation. A practical commentary on

DATA TRANSMISSION – THE PHYSICAL LINK

the recommendation is valuable for anyone designing or using an interface.

The recommendation applies to the interconnection of terminals (data terminal equipment or DTE) and modems (data circuit terminating equipment or DCE). The interconnecting circuits are called interchange circuits and allow the transfer of binary data, control, grounds and timing information.

The recommendation may be applied to:

— synchronous or asynchronous transmission;

— data communications on leased lines, either 2-wire or 4-wire, point-to-point or multipoint;

— data communications on the public switched telephone network (PSTN), either 2-wire or 4-wire;

— links between terminals or between terminals and other devices by using a 'null-modem' interconnecting cable.

V.24 defines two sets of interchange circuits, the widely known 100 series used for the transfer of data, timing and control signals, and the 200 series used for automatic calling. These circuits are described in detail by Scott (1980). A summary of the 100 series circuits with RS-232 equivalents is shown in Table 2.1.

A core of eight circuits in the 100 series can be identified and form the basis for the majority of applications. These are listed in Table 2.2. The other less commonly used circuits are available for more advanced higher speed modems which provide, for example, dual-speed operation and synchronous transmission.

The RS-232 circuits are grouped into ground, data, control and timing signals which are referred to as A, B, C and D circuits. In practical discussion of the interface the signals are usually referred to by acronyms, such as DTR for data terminal ready – these are given in Table 2.1. RS-232 defines interchange circuits which are largely a subset of V.24, but with different names and two circuits not in V.24.

RS-232	V.24	Direction	Pin	Abbrv	Description
AA	191	Bothway	1	PG	Protective Ground
AB	102	Bothway	7	SG	Signal Ground
BA	103	To Modem	2	TD	Transmitted Data
BB	104	From Modem	3	RD	Received Data
CA	105	To Modem	4	RTS	Request to Send
CB	106	From Modem	5	CTS	Clear to Send
CC	107	From Modem	6	DSR	Data Set Ready
CF	109	From Modem	8	CD	Data Channel Received Line Signal Detector, Carrier Detect
CD	108/2	To Modem	20	DTR	Data Terminal Ready
CE	125	From Modem	22	RI	Ring Indicator
	108/1	To Modem		CDSL	Connect Data Set To Line
CG	110	From Modem	21		Signal Quality Detector
CH	111	To Modem	22		Data Signal Rate Detector
CI	112	From Modem	23		Data Signal Rate Detector
DA	113	To Modem	24	ST	Transmitter Signal Timing Element
DB	114	From Modem	15	ST	Transmitter Signal Timing Element
DD	115	From Modem	17	RT	Receiver Signal Timing Element
	128	To Modem		RT	Receiver Signal Timing Element
SBA	118	To Modem	14	STD	Secondary Transmitted Data
SBB	119	From Modem	16	SRD	Secondary Received Data
SCA	120	To Modem	19	SRT	Secondary Request to Send
SCB	121	From Modem	13	SCT	Secondary Clear to Send
SCF	122	From Modem	12	SCD	Secondary Received Line Signal Detector

Table 2.1 Serial Interface Interchange Circuits

DATA TRANSMISSION – THE PHYSICAL LINK 35

V.24 Code	Abbreviation	Definition
102	SG	Signal Ground
103	TD	Transmitted Data
104	RD	Received Data
105	RTS	Request to Send
106	CTS	Clear to Send
107	DSR	Data Set Ready
108/1 or 108/2	DTR	Data Terminal Ready
109	CD	Carrier Detect

Table 2.2 Basic Set of V.24 Interchange Circuits

RS-232 also defines the electrical characteristics of the interface (almost identical to V.28) and the connector configuration, which are not part of V.24.

— CCITT Recommendation V.28 (1972): Electrical characteristics for unbalanced double-current interchange circuits.

Signalling at the V.24 interface is by different voltage levels. A positive voltage represents a 0 or space condition and a negative voltage represents a 1 or mark condition. The magnitude of the voltage can vary between set limits of $-3V$ to $-25V$ and $+3V$ to $+25V$, but levels of $\pm 12V$ are common.

The recommendation begins with a statement that the electrical characteristics generally apply to interchange circuits with data signalling rates below 20,000 bits per second. In practice, a set of preferred bit rates has been widely used. These are 110, 150, 200, 300, 600, 1200, 1800, 2400, 4800, 9600 and 19200. There is nothing in the recommendation that gives any guidance on the bit rate for a particular application.

This is an appropriate moment to consider the difference between bit rate and baud rate, a matter of some confusion. Bit rate refers to the binary information transfer rate, ie the number of 0 and 1 binary digits transmitted in one second. Baud rate is the unit of signalling rate or modulation rate, in transitions per second. Baud rate equals bit rate when two-state signalling is used. Because of baud rate limitations on speech circuits, multi-level signalling is used to obtain higher bit rates. There still remains a grey area in the description of the bit rate for asynchronous transmission. Bit rate should be limited to the information content of the signal, and the start and stop elements do not strictly carry information. The true information rate will therefore be anything up to 25% lower than the quoted 'bit rate'.

A more detailed discussion of V.28 follows with consideration of the speed and distance possible in serial connections.

— International Standard ISO 2110: Data communications – 25-pin DTE/DCE interface connector and pin assignments.

This standard specifies a 25-pin Cannon D-type connector to be used for the interconnecting lead between terminal and modem and gives the pin assignments for V.24 circuits.

MAKING THE CONNECTION

The V.24 standard is designed for data communication over dial-up telephone lines or leased lines. The use of the various circuits will be described for the dial-up link as this illustrates the use of all the commonly used circuits. The steps needed to make a direct connection between two computers can then be derived fairly easily. A complete connection is shown in Figure 2.3.

A data call can be split into two phases in the same way as a telephone call: call establishment and controlled data flow. In a voice telephone call the call establishment part of the call includes the calling party picking up the handset, dialling the number, listening for ringing tone and waiting for the called party to pick up the other handset, which completes the connection. Conversation is then possible and the call is held until the calling party replaces the handset. A data call can be set up in the same way, with an arrangement to switch the telephone line from the handset to the modem when the connection has been made. Two circuits are used

DATA TRANSMISSION – THE PHYSICAL LINK

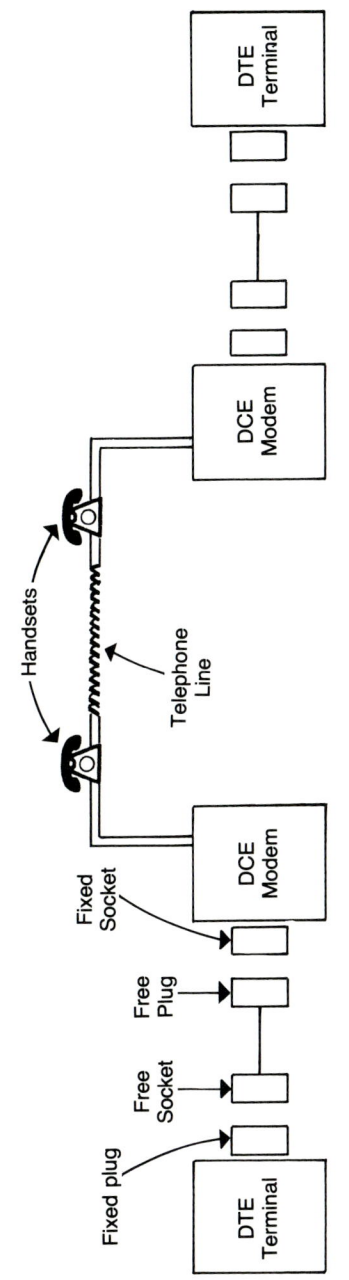

Figure 2.3 The V.24 Interface in a Manually Connected Data Link over the PSTN

in the call establishment phase: data terminal ready (DTR) and data set ready (DSR). The sequence of events in a manually connected call is shown in Figure 2.4.

Assuming that the computer interface is powered up, the line will be high, and will be switched to the modem interface when the data switch is pressed. This may be located on the handset or the modem, but the effect of switching to data is the same. The telephone line is connected to the modem output circuits and DTR high is switched to the modem interface. The modem interface in turn sets DSR high which is returned to the computer interface. This interlocking arrangement occurs at each end and completes the call establishment phase. The diagram illustrates DTR controlled by the handset.

The data transfer can now take place, but first some more terms must be explained. The data transfer may be simplex (a transfer in one direction only), half-duplex (each end can send data but not simultaneously) or full-duplex (each end can send data simultaneously).

Figure 2.4 illustrates the half-duplex mode of data transfer as this requires full use of the V.24 control signals. The circuits used to control the transfer of data are request to send (RTS), clear to send (CTS) and carrier detect (CD). Half-duplex mode of transmission is commonly found in terminal to mainframe connections for reasons of economy and efficiency. The method of flow control is known as polling and is described in more detail in the section on terminal emulators.

In a polled system the mainframe is the master device and controls data transfers to and from the terminal with sequences of control characters. Before any transfer can take place the line must be conditioned: this begins with the computer interface raising RTS. The modem responds to this by turning the carrier signal on, and after a short delay, CTS which is returned to the computer interface. At the far end the other modem detects the carrier and raises CD which prepares the terminal to receive data and also prevents it attempting to send data. The computer then sends the initial data packet, and terminates this transfer by setting RTS to the low condition which removes the carrier from the line. The other modem responds by setting CD low and the process can then

DATA TRANSMISSION – THE PHYSICAL LINK

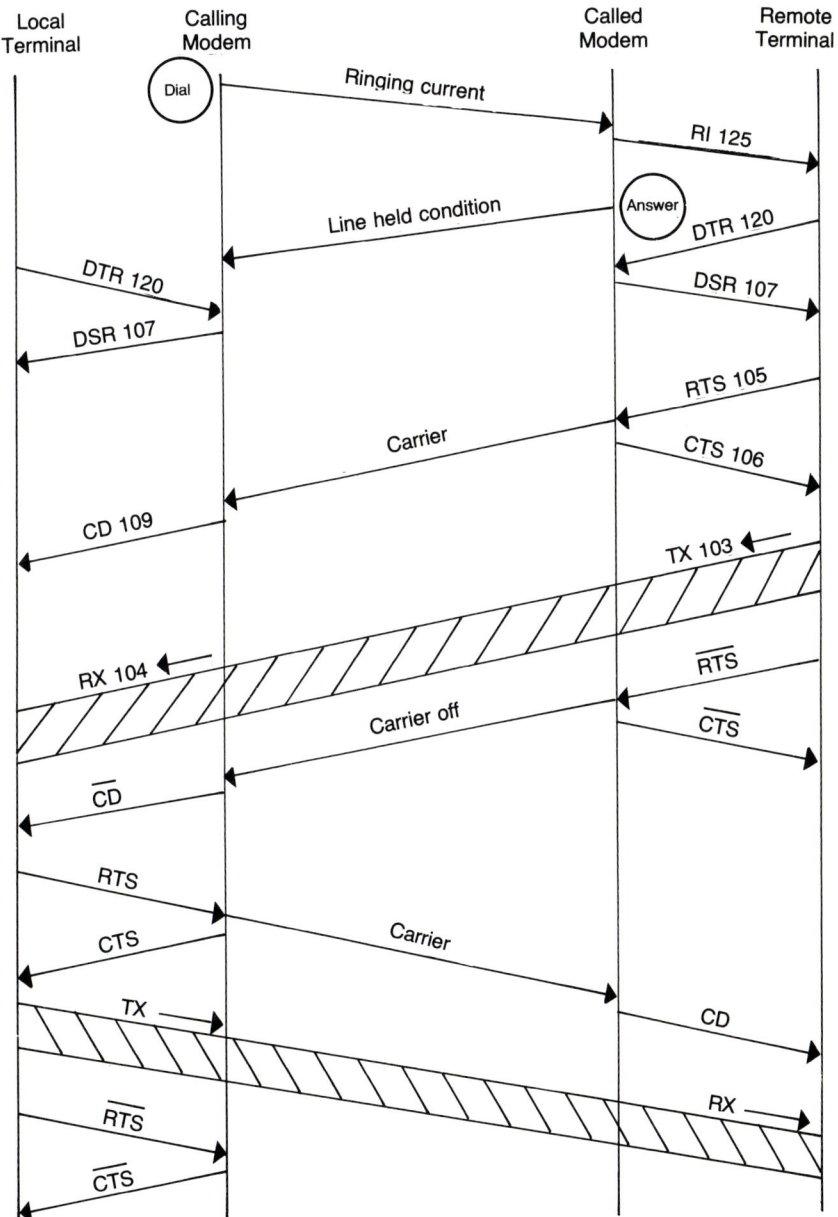

Figure 2.4 Flow Chart for a Half-Duplex Data Link

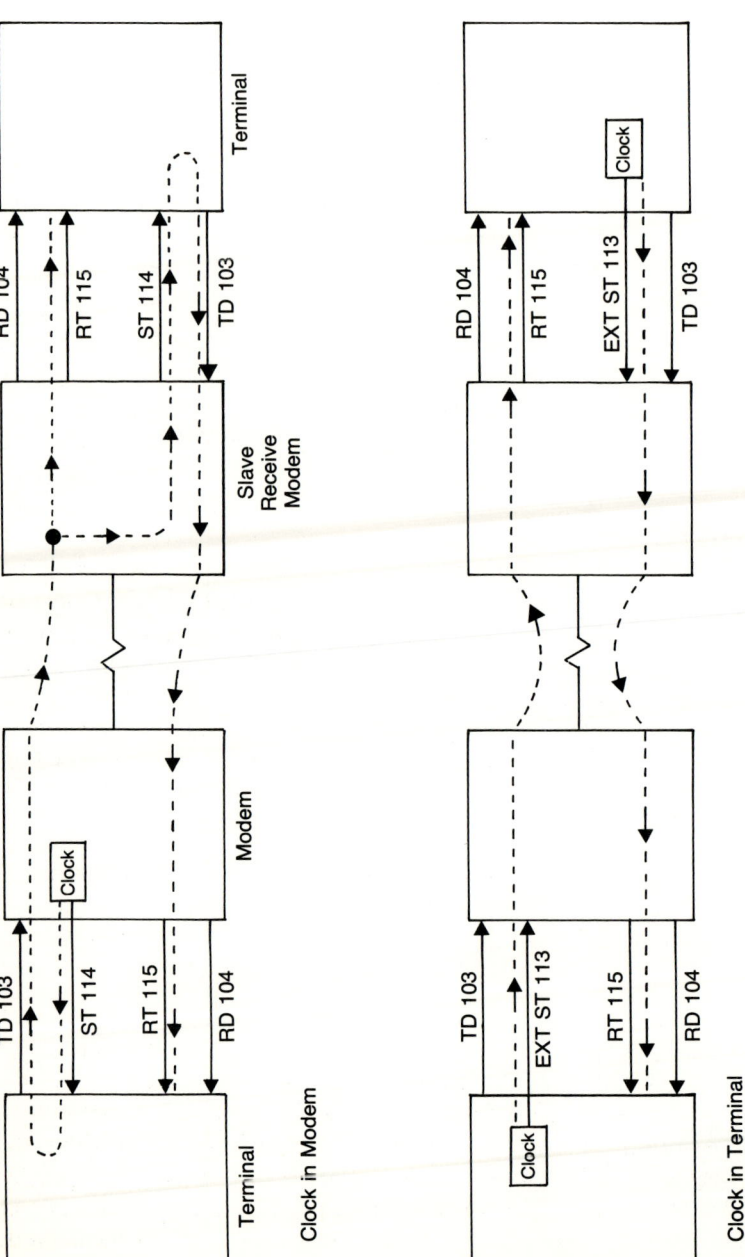

Figure 2.5 Connection of Timing Circuits for Synchronous Links

DATA TRANSMISSION – THE PHYSICAL LINK

continue in the reverse direction.

Data transfer should then take place in an orderly alternating pattern between terminal and computer. Problems of contention for the line, synchronisation and recovery from line faults are the responsibility of the data link protocol. One important point to note about proprietary protocols is that due to their complexity some protocols have been frequently revised which adds to the problems of interworking.

Half-duplex transmission is usually found in high speed synchronous transmission links. Some additional V.24 circuits between terminal and modem are required to complete the synchronous connection. These are the transmit and receive timing signals, which are sent between terminal and modem to synchronise the clocks in the two interfaces. V.24 does not specify which device should be the master, and two timing circuits are defined, both named transmitter signal timing element (ST). The signal on pin 24 is used when the terminal provides the master clock, and the signal on pin 15 is used when the modem is the master device. The two interfaces must be correctly configured for external or internal timing to prevent a conflict occurring. There are in addition two receiver signal element timing signals (RT). V.24 circuit 115 is used when the clock is located in the modems; circuit 128, which is rarely implemented, is used when a terminal is required to clock in data using its own internal timing. ISO 2110 only defines pin 17 for circuit 115.

Figure 2.5 shows the two methods for generating the transmit timing. When the clock is located in the modem it is usual to synchronise the send and receive clocks by deriving the clock in one modem from the incoming signal. This is known as the slave receive modem.

The sequence of events in a full-duplex transmission is shown in Figure 2.6. The same flow control circuits – RTS, CTS and CD – are used, but as there is no need to synchronise alternating data transmission these circuits can be on all the time. The diagram shows RTS and CTS on when the interfaces are powered up. The modem is therefore transmitting carrier continuously and this will be passed to the other modem as soon as the call is established. Both modems will then return CD high to the terminal and compu-

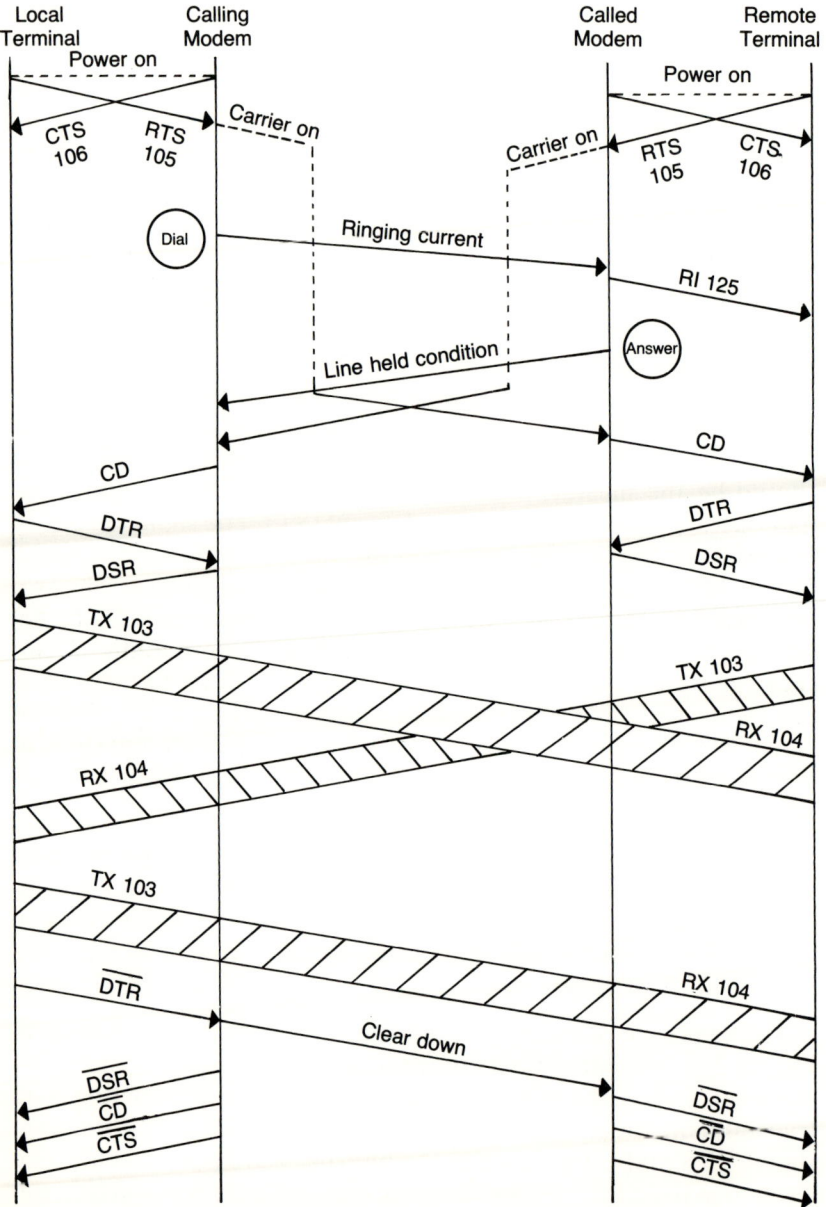

Figure 2.6 Flow Chart for a Full-Duplex Data Link

DATA TRANSMISSION – THE PHYSICAL LINK   43

ter interfaces so that the transmission can begin as soon as DTR and DSR have been set to high. Data can be transmitted simultaneously in each direction until one end lowers DTR to indicate end of transmission.

Full-duplex transmission is the normal mode for asynchronous terminals and in this case the circuits shown in Figure 2.2 are sufficient to control the data transmission. As there is no need to turn RTS and CTS on and off during the transmission these circuits can be provided by passive circuitry.

SUMMARY

An interface described as conforming to V.24 or RS-232 is not a complete description of the interface. This description has become accepted to mean:

- The interface uses a subset of the circuits defined in V.24. The interface may therefore have a mixture of actively- or passively-supported circuits and unsupported circuits.

- The electrical characteristics conform to recommendation V.28. In practice the voltage levels are likely to be $\pm 12V$.

- A 25-way Cannon D-type connector is used, with pin assignments specified by ISO 2110. If the interface is configured as a terminal the connector will be a plug, and if a modem the connection will be a socket.

3 The Serial Interface in Direct Connections

CROSS CONNECTIONS

When a terminal or microcomputer is to be linked directly to a computer the modems and telephone line are removed from the equation, but the serial interface at each end of the link still expects to be talking to a modem. Fortunately, there is a simple solution to this problem as the V.24 circuits can be cross connected in the cable or looped in the connectors to place data, control and timing signals on the right pins as if they had passed through the modem-to-modem link. A cable which provides this function is known as a null-modem cable. Because of the number of circuits involved, there are several variations possible, but the most general solution for half-duplex operation is shown in Figure 3.1.

This set of cross connections links the following circuits:

— TD (pin 2) for each connector is linked to RD (pin 3) of the other connector;

— RTS (pin 4) is looped to CTS (pin 5) in each connector;

— RTS (pin 4) is linked to CD (pin 8) in the other connector;

— DTR (pin 20) is linked to DSR (pin 6) in the other connector;

— ST (pin 25) for each connector is linked to RT (pin 17) of the other connector. ST (pin 15, transmit timing signal from modem) is strapped to ST (pin 24, external timing signal to modem) in each connector;

— ground lines (pins 1 and 7) are directly linked through the cable.

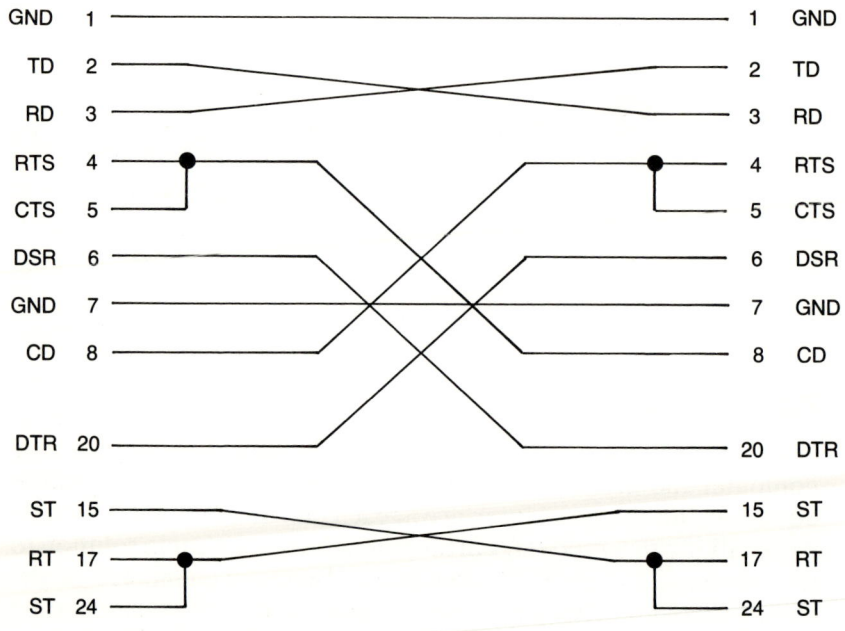

Figure 3.1 Cross Connections for Null-Modem Cable

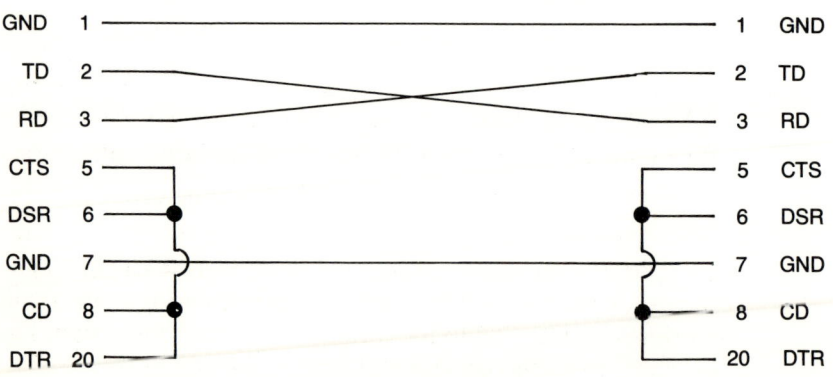

Figure 3.2 Simplified Null-Modem Cable

THE SERIAL INTERFACE IN DIRECT CONNECTIONS 47

This set of cross connections correctly transfers data signals (TD and RD), call establishment signals (DTR and DSR), flow control (RTS, CTS and CD) and timing signals (ST and RT).

A full-duplex asynchronous connection, which is the norm for microcomputer to microcomputer links, can be made with a simplified null-modem cable. This set of cross connections is shown in Figure 3.2. Only ground and data signals are carried by the cable, which is the type of cabling found with asynchronous terminal systems. CTS, DSR, CD and DTR are looped together in each connector, so that DTR, the output signal, enables the interface by raising the other three input circuits high when the interface is powered up. There may be instances where DTR is not provided whilst CTS, DSR and CD are active. In these cases an alternative cross connection is required. This requires a study of the interface manual or the use of a break-out box. This device, essential equipment for the communications specialist, enables all the V.24 circuits to be monitored with LED displays, by placing it in series with the link. Although there are many variations, a most useful break-out box shows the high state as red, the low state as green and the grounded or disconnected circuit by both LEDs off. In addition, each circuit can be broken by a switch and cross connections made between circuits.

Problems with cross connections are due to the fact that commercial implementations rarely use the full subset of V.24 circuits for a particular application. For example, a terminal may require DSR high to enable transmission, but there may be no provision for this signal in the cable or from the remote interface. In such cases, pin 6 at the terminal must be strapped to another pin at the appropriate d.c. level. This can be taken as a general principle: pins which are receiving signals must be strapped to other output pins which have the correct condition for normal operation.

Another hardware variation that may be encountered in some commercial serial interfaces is the modem/terminal switch. This is usually a dil package containing cross connections which is mounted on the serial interface card. As the terminology suggests, this switch block makes the interface look like a modem in one position and a terminal in the other position. The switch block contains cross connections which perform the job of a null-modem cable. A microcomputer with the interface set to the terminal position can then be connected to a second microcomputer with the interface set to the modem position using a cable with straight-through connections.

HOW FAR, HOW FAST?

Having established the correct character format and cross connections required, the next question to consider is what is the maximum baud rate that can be used and over how far. These characteristics of a transmission line are inversely related, and it should be possible to derive a curve of speed against distance as shown in Figure 3.3. There are so many factors involved however, that in practice such a curve is difficult to produce. This problem is discussed in detail by McNamara (1977).

The serial interface was designed for terminal-to-modem connections, which would normally be over relatively short distances.

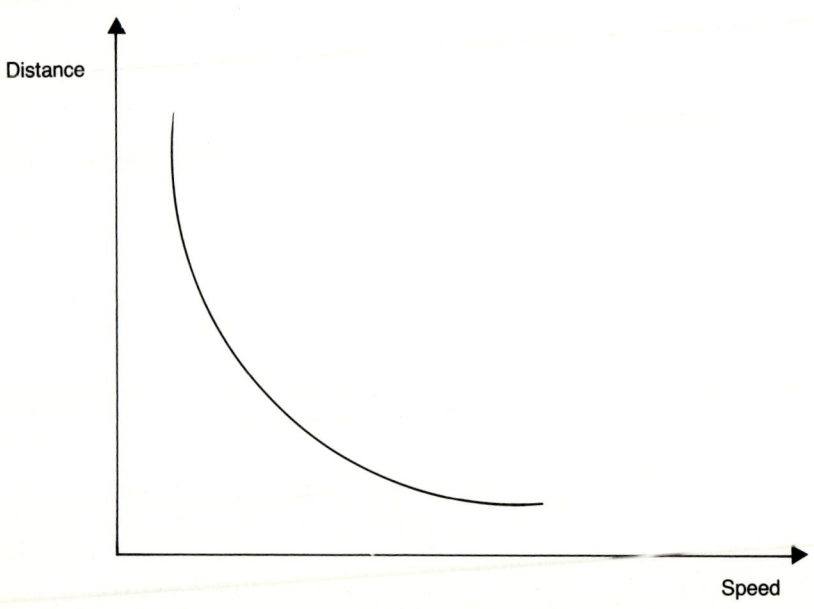

Figure 3.3   Idealised Speed versus Distance Curve

THE SERIAL INTERFACE IN DIRECT CONNECTIONS

RS-232-C sets a maximum interchange circuit length of 50 feet. V.28 does not set a limit, although British Telecom recommends 30 metres for data rates up to 2400 bit/s.

In practice much longer distances are possible with high-quality cable, and it is quite feasible to operate at 1200 bit/s over half a mile. It all depends on the environment, the cable and whether synchronous or asynchronous transmission is used. For long distances the requirements are:

— an electrically 'quiet' environment, to minimise induced currents in the cable which can lead to misinterpretation of signals;

— low-capacitance cable, to minimise pulse degradation;

— asynchronous transmission to avoid cross-talk caused by clock pulses on adjacent circuits.

For on-site operation, line drivers and base band modems are commonly used to overcome the limitations of the V.28 specification.

In order to answer the question of how far and how fast more fully the V.28 recommendation will be examined in more detail. The title of V.28: "Electrical characteristics for unbalanced double current interchange circuits" requires further explanation:

— an interchange circuit has two leads to provide a go and return path for current flows;

— a double-current interchange circuit is one in which current may flow in either direction, depending on the polarity of the voltage applied;

— an unbalanced interchange circuit is one in which one of the leads is at ground (earth) potential.

An electrical model of the complete interchange circuit is shown in Figure 3.4. It consists of a voltage generator, which in the case of the TD circuit would be in the terminal, connected to a load which would be in the modem. The line of demarcation between the generator and load is at the connector attached to the modem. The impedance of the line is therefore lumped together with the impedance of the voltage generator.

When disconnected, the voltage produced by the generator must not exceed 25 volts. When connected to a load having a resistance in the range 3000 to 7000 ohms the voltage must remain between 5 and 15 volts. The capacitance of the load including the line must not exceed 2500 pf.

V.28 therefore places a restriction on line length irrespective of speed on account of the capacitance limit. For a typical cable with a capitance of 150 pf/m, the maximum allowable cable length would be 16 m.

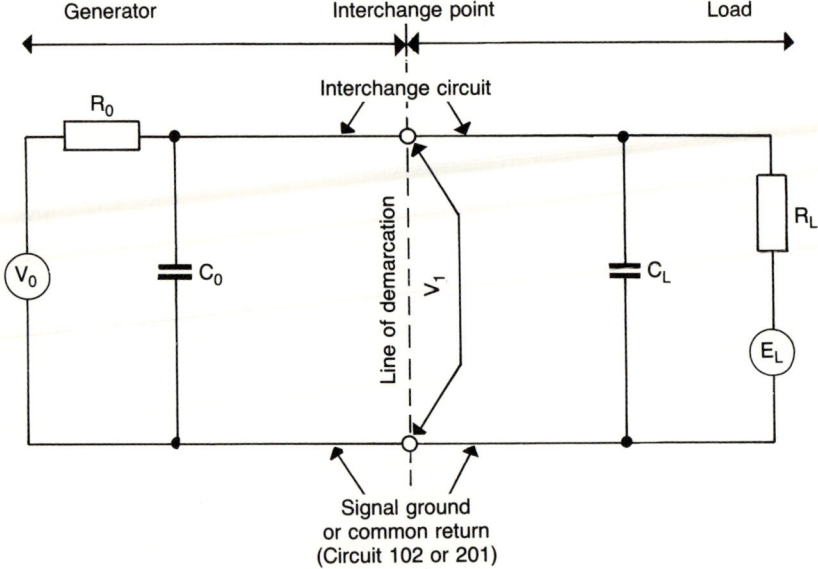

V_0 is the open-circuit generator voltage.
R_0 is the total effective d.c. resistance associated with the generator, measured at the interchange point.
C_0 is the total effective capacitance associated with the generator, measured at the interchange point.
V_1 is the voltage at the interchange point with respect to signal ground or common return.
C_L is the total effective capacitance associated with the load, measured at the interchange point.
R_L is the total effective d.c. resistance associated with the load, measured at the interchange point.
E_L is the open-circuit load voltage (bias).

Figure 3.4 Electrical Model of the V.28 Interchange Circuit
(Reproduced by kind permission of the ITU)

NEW ELECTRICAL STANDARDS

The V.28 standard was developed in the discrete components era, and the electrical characteristics defined are not compatible with current integrated circuit technology. The limited speed and distance possible for interfaces designed to V.28 has often proved to be a handicap.

Two newer standards exist which are an improvement in many ways, in that they are IC-compatible, permit operation over much longer distances and operate at much higher speeds. They originated as EIA standards and are designated RS-423 and RS-422. They were subsequently adopted by CCITT and published as recommendations V.10 and V.11, and for public data networks as X.26 and X.27. A further standard – RS-449 – has been published by EIA to cover the non-electrical aspects of the interface (such as connectors) and this updates RS-232-C.

V.10 is designed for unbalanced interchange circuits and V.11 for balanced interchange circuits. The lower capacitance per unit length of a balanced circuit means that it can be used for longer distances at higher data rates compared with an unbalanced circuit. Interfaces to these standards will replace V.24/V.28 interfaces where higher performance is required.

CCITT Recommendation V.10

V.10 operates at up to 100K bit/s with unbalanced interchange circuits. It uses voltages between ± 3 and ± 6 volts, or if limited interworking is required with V.28, ± 4 volts and ± 6 volts. In each case the threshold voltage is 0.3 volts.

The equivalent circuit for the interface is shown in Figure 3.5 which shows the receiver as a differential receiver, although the line is unbalanced. This allows compatibility with V.11 so that balanced and unbalanced circuits can be mixed in the same interface. In the V.10 interface one of the inputs to the receiver circuit would be grounded and connected to the common return path.

CCITT Recommendation V.11

V.11 operates at up to 10 Mbit/s with balanced interchange circuits. The voltages used are the same as in V.10 and the receiver is

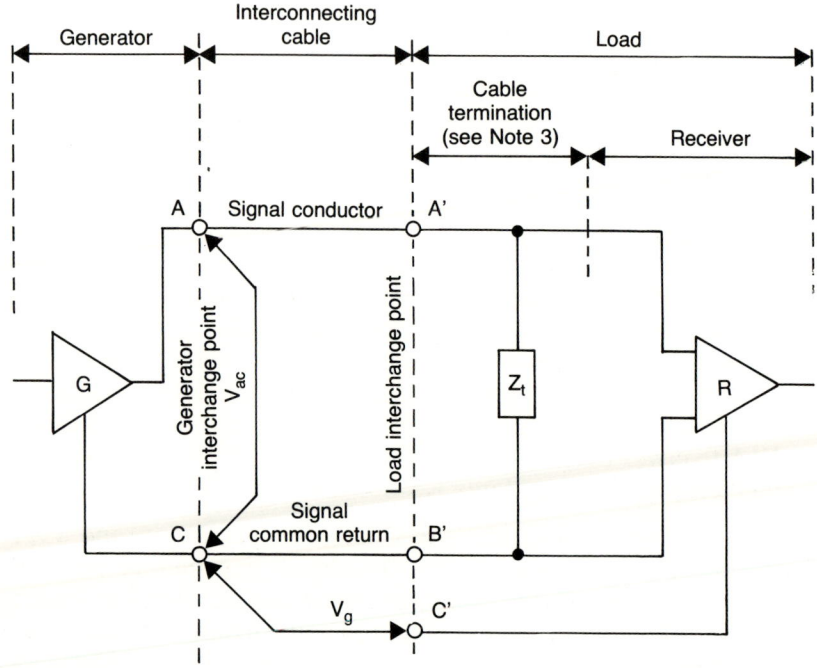

V_{ac} = generator output voltage
V_g = ground potential difference
A = generator active interchange point
C = generator common return point
A' = load active interchange point
B' = load common return point
C' = receiver zero reference point

Note 1 – Two interchange points are shown. The output characteristics of the generator, excluding any interconnecting cable, are defined at the "generator interchange point". The electrical characteristics to which the receiver must respond are defined at the "load interchange point".

Note 2 – Points C and C' may be connected to protective ground if required by national regulations.

Note 3 – The interconnecting cable is normally not terminated.

Figure 3.5 Equivalent Circuit for the V.10 Interchange Circuit
(Reproduced by kind permission of the ITU)

THE SERIAL INTERFACE IN DIRECT CONNECTIONS

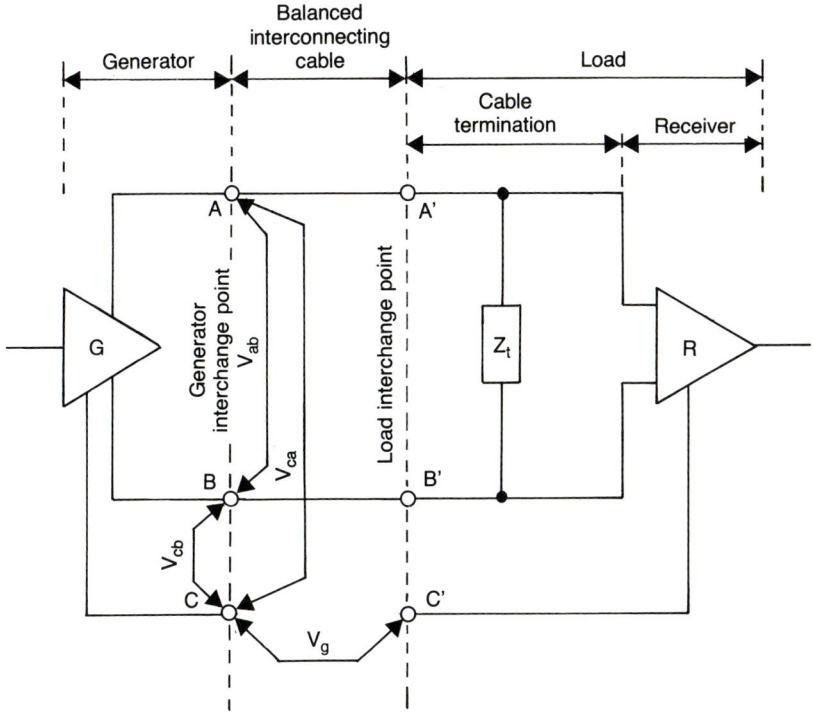

V_{ab} = generator output voltage between points A and B
V'_{ca} = generator voltage between points C and A
V'_{cb} = generator voltage between points C and B
Z_t = cable termination impedance
V_g = ground potential difference
A, B and A', B' = interchange points
C, C' = zero volt reference interchange points

Note 1 – Two interchange points are shown. The output characteristics of the generator, excluding any interconnecting cable, are defined at the "generator interchange point". The electrical characteristics to which the receiver must respond are defined at the "load interchange point".

Note 2 – Points C and C' may be interconnected and further connected to protective ground if required by national regulations.

Figure 3.6   Equivalent Circuit for the V.11 Interchange Circuit
(Reproduced by kind permission of the ITU)

identical to that in V.10. The equivalent circuit is shown in Figure 3.6.

V.10 and V.11 can in many cases operate with links several kilometres long at the lower signalling rates. Choice of which standard to use is broadly determined by speed – up to 20K bit/s use V.10, over 20K bit/s use V.11.

THE OSI SEVEN-LAYER MODEL

A guide to microcomputer communications would not be complete without a discussion of network architectures and in particular the work of the International Standards Organisation (ISO). At the present time ISO are responsible for defining the detailed specifications for a communications architecture, known as the Open Systems Interconnection (OSI) Seven-Layer Model. This set of internationally agreed standards defines a complete set of protocols which will form the basis of a structure for the interworking of different computer products. The architecture follows a layered structure, where each layer uses the services of the lower layer, starting with the physical layer, which provides the transmission line and hardware interface.

When this approach was first applied to data communications the term 'onion skin technique' was used. Data which passes through a communications link is subject to a series of operations, each of which can be considered separately, which are reversed at the destination. Hence the operations were regarded as a series of layers, like an onion, which could be peeled off to reveal the layer below.

ISO has grouped the operations into seven distinct layers as shown in Figure 3.7. This is not the only possible architecture, the main contender being IBM's Systems Network Architecture (SNA) which organises the operations into six layers. Since SNA was first introduced in 1974 it has achieved such widespread success that many other suppliers have adopted it. IBM claim that SNA is also an open system as the protocol definitions are available to other companies. However, the important difference between a proprietary protocol and an international protocol is that the proprietor can update the definitions without reference to any other party.

When the full set of OSI standards has been defined, any two pieces of equipment built to these standards will be able to communicate despite differences in construction, CPU type and power, manufacturer, functional organisation and internal interfaces.

An open system must provide both interconnection and interworking. These two requirements are distinguished by their levels in the hierarchy of protocols. The interconnection protocols begin with the electrical standards for the transmission and continue

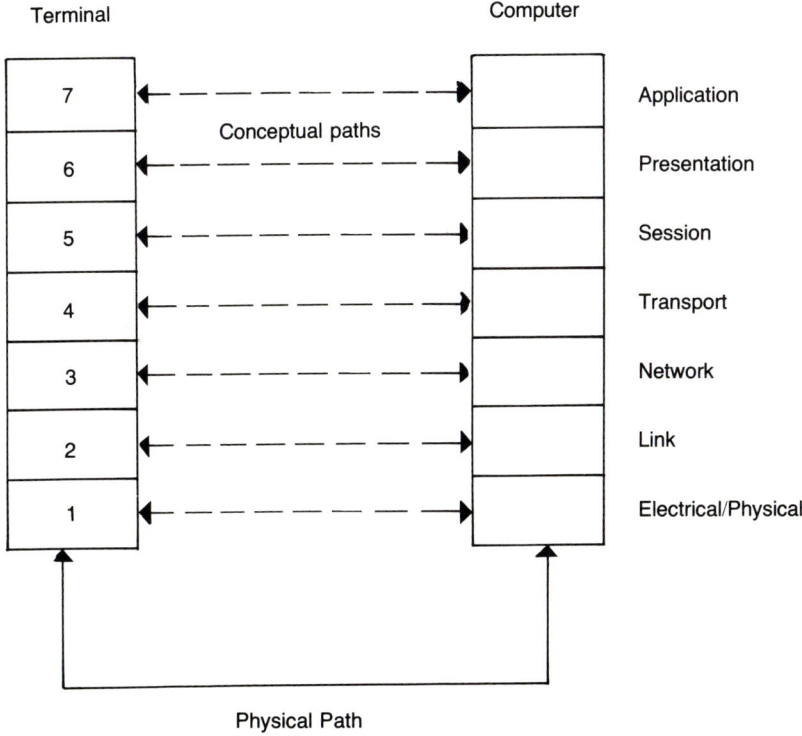

Figure 3.7 OSI 7-Layered Architecture

through error control, interconnection of physical lines to form circuits, through to choice of circuit quality. These low-level protocols are the responsibility of the bottom four OSI layers – the physical, data link, network and transport layers. They are considered to complete the specification of the data communications functions.

The physical layer (1) provides the interface between the computer equipment and the physical line. It is responsible for the control of the flow of bits across this interface. An example of a standard at this level is CCITT Recommendation V.24.

The data link layer (2) groups the bits provided by level 1 into frames or sequences of frames. This layer also controls access to the transmission medium where it is shared by several machines, and can provide error detection and correction. The best known standard at this level is ISO's High Level Data Link Control (HDLC).

The network layer (3) adds extra routeing facilities to enable the frames from level 2 (more commonly called packets) to be passed over a succession of data links within one or more networks. This layer can also provide error control, multiplexing and flow control.

The transport layer (4) provides for communication between processes in the attached equipment. It provides end-to-end flow control of the movement of data packets between these processes. The selection of class and quality of service are also the responsibility of level 4.

Interworking is covered by the three highest levels of the protocol hierarchy – the session, presentation and application layers. These layers are usually considered to be data processing functions.

The session layer (5) establishes logical communications paths between applications or users wishing to exchange information. The dialogue between users is known, in OSI terminology, as a session. The session layer establishes, maintains and breaks the dialogue, and ensures that data reaching a system is routed to the correct application. It ensures that the information exchange is correctly synchronised and delimited so that, for example, two applications do not try to transmit simultaneously.

THE SERIAL INTERFACE IN DIRECT CONNECTIONS 57

The presentation layer (6) is concerned with the interpretation of the data exchange during a dialogue. It is responsible for passing data received to the applications in a form they can understand, and for displaying the data to the user in the manner designed for the application. It therefore includes handling of character, screen format and graphics codes, and may employ encryption techniques. By providing services independent of internal character formats it gives machine independence.

The applications layer (7), the highest level defined, provides services to applications wishing to communicate. These services include:

— file transfer, access and management;

— job transfer and management;

— virtual terminals;

— common application service elements, eg checkpointing and bulk transfer;

— control of application process groups.

A particular communications requirement will need a subset of the seven layers, which can be implemented in a variety of ways. In a microcomputer, for example, all the required layers may be provided in a single program, whereas in a corporate network containing large host computers, the layers may be split between programs and between processors.

The OSI model has proved to be a valuable aid in understanding communications software, and its importance can be judged by the number of companies engaged in implementing the protocols already defined.

4 Terminal Emulation

WHY TERMINAL EMULATION?

The decision about whether to use a microcomputer with terminal emulation and how to select a terminal emulation depends on the type of user you are. You might be considering terminal emulation because:

— you have a microcomputer and are a casual user of mainframe or minicomputer systems, but you do not have easy access to a terminal;

— you want to provide basic terminal facilities at the lowest possible cost;

— you are in a computer support group and you want a single terminal able to access a number of computer systems from different manufacturers. The original terminals are likely to be sophisticated terminals, connected to the host computer by a synchronous link;

— you are in a DP department and you want to install a cluster of terminals with enhanced features not found on the original terminals.

Whilst the casual user might be satisfied with a home computer fitted with a basic emulator chip running at 300 baud, the dp professional would look for a purpose-built multicomputer system giving complete emulations with additional capabilities.

The communications needs of the microcomputer user can be grouped into three categories:

- 'glass teletype' emulation. This basic communications facility allows commands to be sent to the remote computer and responses to be displayed. Control codes can be sent to the computer and responses transferred to a scrolled display, simulating the paper printout of an electromechanical teletype;
- sophisticated terminal emulations, with file transfer. Additional display facilities such as direct cursor addressing will be supported. These may be proprietary terminal features as may be offered by, for example, a DEC VT52 or VT100 terminal, and may include a widely accepted standard set to the ANSI specification. File transfer will be provided for specific target host computers;
- fully integrated microcomputer to host computer file transfer. Products in this category are designed for commercial users of data base systems. Local applications are set up on the microcomputer, using spreadsheet, word processing or data base software, but with data supplied by an enquiry to the host computer data base. The communications software takes care of the commands to the host computer operating system.

The performance and limitations of terminal emulators will be examined in the next section as a guide for prospective users. Product reviews will be kept to a minimum because of rapid developments and growth in new product announcements, but some evaluations are useful as a guide on what to look out for. A current product survey will be maintained by the NCC Communications Division in collaboration with the Microsystems Centre.

CHOOSING A TERMINAL EMULATION

It is an elementary task to transfer characters between a microcomputer keyboard and a host, and conversely between the host and microcomputer display screen. Microcomputer operating systems provide I/O device drivers which may be called from a Basic program. Using these routines, a skeleton terminal emulator can be written in just a few lines of Basic.

It will be restricted in speed and will not provide any special display features, but will allow commands to be sent to a remote host and the output to be displayed.

The software required becomes progressively more complex as more facilities are added and higher speeds are aimed for. Timing becomes critical as host computer control sequences are incorporated (usually Escape and several other characters) which need to be interpreted and acted upon. Further problems arise with synchronous terminal emulation, which is described later. Because of the performance limitations of a general-purpose microcomputer compared with a custom built terminal, features may be deliberately omitted. It is therefore important to check on which original terminal features are not supported as your application may depend on them.

The kind of features in question are direct cursor addressing (the movement of the cursor to any part of the screen), various screen clear functions (clear entire screen, clear entire screen from cursor, clear to end of line, etc), highlighting features – the list is quite extensive.

As the number of features is extended the question of compatibility with the original terminal arises. The lack of conformance to a character standard has already been mentioned. Even when the microcomputer and host computer use the same ASCII character set, the use of the 32 control characters can be different in several instances. For example, does the microcomputer respond to the TAB character (ASCII 9) by moving the cursor eight spaces to the right? Other problem characters are Delete (ASCII 127) and Carriage Return (ASCII 13). The original terminal may respond to Delete by carrying out a sequence of steps: Backspace-Space-Backspace, but the microcomputer may respond differently. The original terminal may append Linefeed to Carriage Return, the microcomputer may pass Carriage Return unmodified.

Terminal emulations are therefore found for specific host computers, so that the use of control characters and escape sequences match those of the original terminal. The conversion of control sequences may be conveniently provided by look-up tables which convert the host control sequences to the required internal codes and vice versa.

SOME TERMINAL EMULATION PRODUCTS

Two products are described: one is a general-purpose asynchronous terminal, and the other emulates a specific synchronous terminal. The points raised may help you to compile a checklist of requirements and may also help in the selection process.

One microcomputer, the BBC Micro, has proved to be very popular in the UK as a low-cost terminal. The combination of quality keyboard, good graphics and integral V.24 serial interface allows the machine to be programmed as a versatile terminal with powerful communications facilities. The emulator software can be written to a read-only-memory (ROM) chip, which is then permanently installed in one of the spare sockets. The capacity of a single ROM is large enough for several different emulations, so that when the software is loaded by a simple keystroke sequence, a menu of different emulations can be offered.

The Workstation

This is a ROM developed by the University of Sussex for the BBC microcomputer. The action at switch-on depends on the position of the socket used for the ROM, one position resulting in the Workstation software being entered directly. The menu displayed offers the following options:

— set up terminal;

— Tektronix 4010 graphics emulation;

— DEC VT52 emulation;

— file archive/restore.

The first option allows the user to set up the correct baud rate in the serial interface.

The second option offers the widely used Tektronix graphics terminal functions. There are limitations in the BBC Micro design which mean that a complete emulation of the 4010 is not possible. The most fundamental limitation is on the resolution of the screen, which is restricted to 256 by 640 pixels. The 4010 offers 1024 x 780 which gives excellent resolution but at a much higher basic price.

TERMINAL EMULATION 63

The third option is the DEC VT52 emulation which has been recommended by experienced DEC users. Workstation was primarily designed for use with DEC machines, and operation of the VT52 emulation is particularly smooth. By the combined use of shift and function keys all the DEC editing functions are available.

The final option is file transfer, which offers limited file transfer facilities. A detailed discussion of file transfer is left for the next chapter. In summary, the BBC Micro can be fitted with a multifunction ROM which offers a number of sophisticated terminals operating over an asynchronous transmission link. The performance will be adequate for all except the most demanding graphics applications.

Ultracom 1

This emulation is an ICL synchronous terminal emulation designed for Apple II and Apple Lisa microcomputers by Ultracomp Limited. This emulation requires an additional interface card to be installed in one of the Apple peripheral slots. The software has been designed for the Multicom card supplied by Owl Micro-Communications Limited, a multifunction communications card which offers synchronous transmission.

The synchronous terminals operating the ICL CO3 (or full XBM) protocol have several fundamental differences from the asynchronous terminals commonly connected to minicomputers. The obvious difference is that the transmission is synchronous, so characters are not separated by start and stop bits. Clock signals are exchanged between terminal and computer, or terminal and modem, and transmission must be synchronised in a start-up procedure for both ends to be brought in step. But the elimination of start and stop bits only gives a 20% improvement in throughput which is often insignificant in a typically 'bursty' communication. Only when large files are being transferred does this offer real savings, and then only if there are no other overheads.

More significant differences are the organisation of mainframe terminals into clusters with transmission in half-duplex mode, controlled by mainframe polling. The control of data transfers by polling puts additional restraints on the emulator software design, as the orderly process of polling and response must be maintained. If the terminal does not respond correctly, unpredictable results can occur, giving rise to difficulties in logging on and slow response times as the mainframe keeps attempting to get the correct response sequence. The timings are critical and the microcomputer cannot ignore polls when doing something else such as disk access. One consequence of this in the Ultracom 1 emulation is that the mainframe must be set up to have a no-response time-out of at least two seconds, which is longer than required for original terminals.

Mainframe terminals were designed for data entry via formatted screens, rather than the scrolled screens used in 'glass teletype' emulations. This is achieved with a number of terminal features such as highlighted areas, protected fields and various screen editing functions. Protected fields allow a screen format to be set up where field names are stored as labels which cannot be overwritten or deleted, and blank (unprotected) fields for data entry. Data can only be entered in unprotected fields, and a single key is used to move from one field to the next. Screen editing functions allow all unprotected fields to be cleared leaving just the field names in the protected fields. Single unprotected fields can be cleared to the end of a line or to the end of the message.

The message is defined as the characters between a visible start of message indicator and the cursor. Within an unprotected field, lines can be deleted or inserted with no effect on the format of the rest of the screen. All of these features make data entry on a forms basis very simple. When the screen is completely filled with data, the send key is pressed and the entire screen is transmitted after the poll is received. There is no transmission of data until a request is sent, but the interface must continuously monitor the line to receive polls and send the correct responses.

The Ultracom 1 emulation supports most of the screen features of the original terminal, but some are not supported, most noticeably those concerned with clearing and deleting unprotected fields. The clear screen and erase line/message functions not only clear the unprotected fields but also the protected fields, so that field names will be deleted as well as data entered. Insert and delete line functions also take no account of protected fields and change the format of the screen if used.

Another limitation of the emulation is that the screen is limited to 24 lines whereas the original terminal uses 25 lines. The organisation of the screen display into rows and columns is a hardware design feature and there are no display cards available for the Apple II which offer 80 columns by 25 lines. Ultracom I overcomes this limitation by providing a 24-line window into the 25-line screen. By scrolling up or down the unseen line can be recovered. One consequence of this is that when a 25-line screen is transferred to the terminal the top line is left off, which can be somewhat disconcerting until you become familiar with it.

All the other ICL terminal features worked smoothly and are very well documented. Installation is very straightforward, and a detailed knowledge of the communications card is not required as any required configuration is entered via simple-to-use menus. Different configurations containing address data for polling can be stored on the emulator disk and recalled using the system menus.

The Apple Lisa version of the Ultracom software offers more interesting possibilities, and users at the North West Regional Health Authority have shown how the system can be used for data capture. Ultracom emulations do not yet offer file transfer but the Lisa multi-tasking operating system allows screen-based data capture. This, combined with the advanced Lisa software, has enabled the Health Authority to develop a useful distributed data processing system.

An emulation of a sophisticated synchronous terminal based on a general-purpose microcomputer will result in some compromises, and is more likely to interest the existing microcomputer user rather than the dp professional. There are custom-built microcomputers which have been designed to emulate the original terminal very closely, but with the bonus of additional facilities. An example of a machine in this category is the Dacoll Taskmaster which is a multicomputer system offering a complete emulation of the original terminal but with a flexibility not found in the original terminal. Although this machine can be used for other purposes, including emulation of an IBM 3270 and other mainframe terminals, it is most likely to be found in a dp department or a software support group. Software support managers demand terminals that match their customers' terminals exactly, and a machine that can offer several different terminals within one box is an attractive proposition.

The mainframe suppliers have now moved into the intelligent terminal market themselves, prompted by the success of terminal emulations from their competitors. Most notable are IBM with their 3270 PC, and ICL with their Concurrent CP/M range of microcomputers.

5 File Transfer

INTRODUCTION

Whilst a terminal emulation package can be a valuable asset to the microcomputer user, it will not be long before the need for comprehensive file transfer facilities arises. A terminal emulator is limited to the display of text files on the screen using the host computer editor or other utilities. Binary files may be examined if a dump utility is provided which displays binary values in hexadecimal form with ASCII equivalents. Files cannot however be transferred to disk as they are being viewed. Early attempts to transfer files from one machine to another used simple programs with rudimentary facilities and often made use of the editor or other utility in the host computer. The programs were system-dependent and therefore difficult to transfer to other computers and were limited in performance. The position has now changed dramatically and a large number of software houses have introduced powerful and sophisticated communications packages. This rapidly expanding market has resulted from both the interest microcomputer users have in communicating with each other and with public databases, and from the introduction of personal computers in industry on a large scale.

FILE TRANSFER – THE PROBLEM OF INCOMPATIBILITY

Why is file transfer between machines so often fraught with difficulties? The commonest type of file to be transferred is a file of characters such as stock levels or sales figures and would appear to be reasonably standard. Unfortunately, the computer suppliers have developed their hardware and software independently and over a very short timescale and standards are slow to be adopted. This problem is aggravated by the desire of suppliers to implement their own proprietary architectures as this helps to lock-in customers and prevent competitors getting a foothold in the market. The main differences which affect file transfer are as follows:

(a) Character sets

The microcomputer user might be forgiven for thinking that there is only one character set in use. This is the 7-bit code, commonly referred to as ASCII, which seems to be used universally on microcomputers. But there are other codes in use, principally the eight bit EBCDIC code used by IBM for their mainframes. As IBM is the largest supplier by far, with approximately 50% of the mainframe market, anything sold by them becomes a de facto standard.

Even ASCII in its international form, ISO-646, presents problems as 12 characters are reserved for national variations, and the 32 control codes are not used consistently between suppliers. There is also considerable pressure for extending the 7-bit code to 8 bits to give an expanded character set, particularly for graphics characters.

(b) File structure

Computer suppliers have constructed a wide range of file structures ranging from simple to complex. The simplest file is a file of characters which is stored exactly as it would be printed, containing only the minimum of control characters, carriage return and line feed, to format the text. More complex files are organised into records containing fields, and have a two-dimensional grid structure. This is to enable sophisticated access methods to be incorporated in the filing system or data base manager as it is usually called. A simple general-purpose file transfer program would only be expected to transfer the file's name and contents but not all the attributes the file possesses. To illustrate the complexity of a computer network which passes file attributes, DECnet supports the following:

FILE TRANSFER

42 system features (random access, sequential access);

8 data types (ASCII, EBCDIC, executable, etc);

4 organisations (sequential, relative, indexed, hashed);

5 record formats (fixed, variable, etc);

8 record attributes (for format control);

14 file allocation attributes (byte size, record size, block size, etc);

28 access options (supersede, update, append, rewind, etc);

26 device characteristics (terminal, directory structured, shared, spooled, etc).

These parameters were included in the communications software package for the family of machines from a single vendor, and are in addition to the more widely recognised attributes such as name, length and creation date. Network services which offer file transfer facilities for machines from different vendors usually make a number of simplifying assumptions and set restrictions in their file transfer protocols. Even with a restricted set of file attributes, many simple operating systems cannot support all the attributes offered, and a file transferred from one machine may be unusable on another.

If a file can be transferred from one machine to another and then back again, and all its attributes recovered the transfer is said to be invertible. Providing invertible file transfers between machines is a major problem with no simple solutions.

Transferring files containing only text information, such as documents, program sources or messages, presents less serious problems as the files are sequential in organisation and require only the simplest of transformations (eg EBCDIC to ASCII).

If an invertible transfer cannot be made between machines by passing attributes, the second machine can still provide a useful service of storage if the attributes are sent as part of the file. The second machine can only perform the temporary storage function, but the first machine can recover the file attributes from the file header when the file is returned.

(c) Operating system

Operating systems differ in their response to characters as they are received or transmitted at the terminal port. All hosts and communications processors pass any printable character to the application program, although the character may need to be converted to a different character coding (EBCDIC to ASCII for example). The response to the 32 control characters is however quite varied between systems. Some control characters may be ignored, some passed to the application and others cause the system to take special action. Some characters (typically DC1 and DC3) are used for flow control and will be absorbed by the I/O driver routine, as will padding characters (NUL or DEL) and the transfer of control character (ESCAPE).

Characters may be passed to the application program immediately or may be buffered by the operating system until a logical record delimiter is received. A logical record is usually a string of characters terminated by a carriage return or line feed. The treatment of the terminator varies from machine to machine. Some, like the IBM VM/370, discard the terminator, whilst others keep it. Different methods of retaining the terminator are used also. For example, UNIX translates a carriage return to a line feed whereas most DEC operating systems keep the carriage return but add a line feed.

In order to transfer a file which contains control characters or 8-bit non-character data the file transfer protocol must translate these characters and data words to printable characters. There are many possible solutions to designing a data transparent protocol and some popular protocols will be described in a later section. Proprietary protocols, such as IBM 2780, also offer data transparency and are widely used by non-IBM vendors because of the strong market position held by IBM.

FILE TRANSFER – TRANSMISSION PROBLEMS

In addition to the incompatibilities between character codes, file structures and operating systems, a number of other problems must be faced during the file transfer. Any file transfers made over the speech telephone network are very prone to errors. All types of noise interference seem possible in a typical telephone circuit, although noise spikes resulting from electromechanical switching equipment are commonly experienced. These result in bursts of data errors which make the transfer of all but the shortest files difficult to transfer error-free. The data-link protocol achieves error-free transmission by dividing the data stream into blocks of characters, or packets, and adding an error checking code to each packet. The receiving protocol can check the received data against the error-checking code and ask for a retransmission if a corruption is found.

During a long file transfer a less powerful microcomputer will not be able to transfer blocks of characters from memory to disk and monitor the terminal port simultaneously. This is because the I/O system may not be interrupt driven, or the interrupt system is not flexible enough to handle disk and I/O interrupting. The solution to this problem, which applies only to the asynchronous link, is to use a flow control protocol. A pair of characters is used to suspend and restore transmission of a file. When the microcomputer receiving the file is close to filling the memory buffer, the character DC3 (CONTROL/S) is sent to the remote machine to inhibit further transmission. When the buffer has been cleared, the character DC1 (CONTROL/Q) is sent to request transmission to be continued.

In synchronous communications links the transfer of data is controlled by the polling mechanism, but there is still a problem for the less powerful microcomputer as the mainframe dictates the timing of the polls. If the microcomputer does not have an interrupt-driven I/O system the polls will be ignored if the CPU is tied up with memory to disk transfers.

A further problem with flow control can be found with some mainframe asynchronous links. As flow control is not required with VDU terminals, some machines are not provided with DC1/DC3 flow control (eg Cyber).

Restrictions on the size of files that can be transferred may also be encountered. Here a distinction must be drawn between microcomputer-to-microcomputer file transfers, which can each be running compatible file transfer software, and microcomputer-to-host computer transfers which make use of the host computer software utilities. In the latter case the file size that can be transferred may be limited by the host editor if that is the utility used, or by the microcomputer memory size if terminal I/O and memory-to-disk transfers cannot be handled concurrently. Any microcomputer file transfer software which makes use of host computer utilities is bound to have many limitations, and comprehensive and flexible file transfer requires a file transfer application to be running in the host computer.

FILE TRANSFER PRODUCTS – A CLASSIFICATION

It is useful at this stage to attempt to categorise file transfer software. The broadest classification that can be made is on the complexity of the software. Although there are software products on the market with all shades of complexity and power, a three-level division has been chosen:

1. Simple: use of supplied utilities, with considerable user effort required.
2. General-purpose communications software: more facilities with better user interface.
3. Interactive communications software: provides a window into a host data base without the user needing a knowledge of the host operating system. Good user interface.

The second classification is made on the data link protocol used, which has two categories:

1. Asynchronous.
2. Synchronous.

The third classification draws a distinction between special file transfer software running in each machine, or at one end only.

1. Terminal emulator with file transfer function.
2. Compatible file transfer software running in each machine.

FILE TRANSFER

There are overlaps between these categories but they do provide a useful framework for discussing communications software products. As file transfer software has matured the list of essential and desirable features has grown considerably. These features include:

1. A menu-driven user interface. This provides a clear and concise way of setting up the parameters in a connection. Cursor control with special function keys and overlapping menus in a window format make for easy setting up.

2. Help screens. A help screen containing an explanation of the menu in use should be available. This saves time looking through the manual.

3. Macros. These are preprogrammed configurations that provide a complete set of parameters for particular connections. This feature can save a great deal of time. Macros can also be provided for transfer to the remote machine to initialise the system and perform housekeeping tasks.

4. Data link protocol. File transfers over the analogue telephone network are subject to errors, and all but the most trivial electronic messages require an error correcting protocol. A choice of protocols is desirable so that the most economical and efficient method can be used for a particular application.

5. Concurrency. File transfer software which works under a concurrent operating system is attractive as it allows messages (eg electronic mail) to be received whilst the microcomputer is being used for another job or jobs. This feature is essential for the new generation of office workstations.

6. Simple built-in editor. A file transfer package which includes an editor allows you to edit messages without leaving the program – another time saving feature.

7. Auto dial/auto answer. This feature provides a software interface to a compatible auto dial/auto answer modem for unattended call connection. With this facility, and a clock facility, the software can be set up to make a call during the cheap period and transfer or receive files from another system. Another possibility is short-code dialling, which allows the user to make a call by entering for example the called party's initials.

8   Clock. With a system clock running, actions can be triggered at present times. This makes the unattended dialling facility possible. It also allows all received messages to be logged with time and date.

9   File management. This feature provides file management facilities within the file transfer program. The features available may include file deletion, renaming, directory listing and various sorting routines.

10 Printer support. This feature allows information sent from a remote system to be dumped onto a printer.

11 Multiple file transfer. By including a 'wild card' character in the name of a file to be transferred, any file which meets the name criterion will be transferred. For example, if * is the wild card for a complete name or extension, *.* will cause all files with all extensions to be transferred. Similarly, if ? is the character wild card, ABC???.XYZ will cause all files which begin with ABC and have the extension XYZ to be transferred.

12 File transfer timing. This is another clock feature which shows approximately how long it will take to transfer a file, and how much time has elapsed since the transfer began.

13 Large screen buffer. When a file is being transferred it is often useful to halt the transfer and look back over the file. By incorporating a screen buffer larger than the usual 24 or 25 lines, it is possible to scroll back over a number of pages.

14 Chat. A chat mode allows two users to transfer text typed on their keyboard to the remote user's screen.

15 Wide availability. This allows interworking between different machines ahead of implementation of file transfer standards.

16 Good manual. With so many diverse features a good manual is essential. A file transfer program can easily approach the complexity of the operating system and a good reference source helps to make the best use of the available features.

17 Monitoring window. Whilst a file is being transferred to disk, the text is displayed on a screen window for an immediate display of the file contents.

FILE TRANSFER

18 Encryption. A high-level feature which will be essential for security conscious staff. Application areas include medical data, personnel files and any other sensitive data.

FILE TRANSFER USING SYSTEM UTILITIES

Before moving on to file transfer products the use of existing system utilities as supplied with the microcomputer operating system for simple file transfers will be examined. The technique relies on the use of utilities such as PIP (peripheral interchange program), the system editor or a dump utility. The method will be illustrated by looking at the CP/M operating system which is widely used by eight-bit microcomputers. This technique has been possible from the time of introduction of the CP/M operating system, and predates custom file transfer products by several years.

The first step is to set up the I/O ports on the two machines to the correct logical devices. There are four logical devices in CP/M:

CON: enter commands from the keyboard and display information on the screen – the operator console function.

RDR: receive information – the paper tape reader function.

PUN: send information – the paper tape punch function.

LST: list information – the printer function.

There are also 12 physical devices allowable. These devices are I/O ports which are accessed by I/O drivers. The I/O drivers include:

TTY: slow console.

PTR: paper tape reader.

PTP: paper tape punch.

These are the usual physical devices used for file transfer although others may be used. The assignment of logical device to physical device is made using a STAT command.

A> STAT   RDR: = PTR:

A> STAT   PUN: = PTP:

The commands set up the paper tape reader as the logical reader device, and the paper tape punch as the logical punch device. The use of the paper tape terminology is a hangover from an earlier generation of machines, but these I/O drivers are ideal for general I/O use and are frequently connected to a modem. It is then only necessary to make the physical connection using a null-modem cable, and to set the baud rate and data formats to be the same in each interface.

The receiving machine is first set to receive a file using the PIP command:

A> PIP   A: MYFILE = RDR: <CR>

The sending machine is then set to transmit the desired file using the PIP command:

A> PIP   PUN: = A: MYFILE <CR>

The file MYFILE, residing on volume A: of the source machine, will be transferred to a file also called MYFILE on volume A: of the second machine. One assumption has been made, that the data link has flow control working. At this point a word of caution is needed. The I/O part of the CP/M system (BIOS in Digital Research terminology) is not supplied by Digital Research, but is customised by the microcomputer supplier. The I/O facilities offered by different suppliers vary in their scope and performance, and some I/O drivers do not include X-ON/X-OFF (DC1/DC3) flow control. Sometimes a suitable I/O driver is available, but cannot be attached to the serial interface port (eg the Apple II with 80 column card). In these cases some patching of the I/O drivers or the CP/M BIOS interface is required.

The previous method would not work for binary files as not only would the operating system absorb some control characters, but I/O drivers such as RDR zero the high bit of each byte transmitted. A solution is available to CP/M users however, as there is a supplied facility called DUMP which converts a binary file into its ASCII hex representation. The file can be sent to the serial port by assigning the logical device LST: to the physical device PUN:. A file can then be sent to the serial port by toggling control P (to cause screen dump to be sent to printer) and then typing:

A> DUMP   A: MYFILE.

The file should then be transferred to the second machine in hex format, and must be recovered using a conversion program such as UNDUMP, which is available through the CP/M user group. UNDUMP leaves the file in a memory buffer, which can be transferred to disk using a SAVE command:

A> SAVE   file size   MYFILE.

Files can be transferred between different machines running different operating systems if the basic utilities are available. The method is only suitable for text files over short distances as file attributes are not transferred and no error correction is applied. A number of points must be checked – these were highlighted in the discussion of the CP/M system:

— physical link compatibility:

　— null-modem cable;

　— baud rates correct;

　— data format consistent;

— correct logical to physical assignments made;

— control codes may cause system action;

— flow control operating;

— manual error checking required;

— correct end of file marker supplied;

— file size may be limited to memory buffer.

In a CP/M system, failure to send the end of file marker, control Z, causes the system to hang indefinitely. If the source file does not terminate with control Z, the assignments on the receiving machine can usually be configured to allow the terminator to be supplied from the keyboard.

FILE TRANSFER PRODUCTS

As communications software is rapidly developing, and the number of new products growing rapidly, a complete product review is likely to become out-of-date very quickly. An up-to-date software product directory is maintained by the Microsystems Centre to cater for this need. A comparison of some existing products is useful however in assessing new products and for giving guidance in defining the facilities you require in your communications software.

Two products which have been available for a relatively long time are BSTAM and BSTMS from Byrom Software. The first product, BSTAM (Byrom Software Telecommunications Access Method), is for file transfer between CP/M microcomputers. The second, BSTMS (Byrom Software Terminal Monitor System), is a terminal emulator for host computers, with a file transfer mode.

BSTAM is a general purpose asynchronous file transfer package for microcomputers running the CP/M operating system or some derivative of it. There are two parts of the software, RECEIVE.COM and TRANSMIT.COM, which must be running in each machine respectively. BSTAM requires patching to configure the serial interface port for a particular machine, and so does not have easy to use menu driven installation features. File names are transferred as well as file contents, and multiple file transfers are possible using the same wild card structure as the CP/M PIP utility. Error checking and retransmission are built into the software and recovery from various line faults is provided.

BSTAM is a basic file transfer package offering error free transfers with few frills. If properly installed its performance can be quite satisfactory, but it may not suit the more adventurous users.

Although the names are similar, BSTMS is a quite different product. It was designed for users of CP/M computers who wanted to use their microcomputer as a terminal to a host computer. The emulation is a simple teletype like terminal for asynchronous links. The main difference between BSTMS and a real teletype being the ability to transfer files onto disk as they are being listed on the screen. The operation of BSTMS as a host computer terminal does mean however that the file transfer method is quite different to that used by BSTAM for file transfers between two CP/M machines.

The reason for this difference is that BSTMS behaves as a terminal in all operations, and provides a means for file transfer during normal terminal operations. Also no assumptions are made about the host computers, so that no special file transfer software is expected to reside on the host.

FILE TRANSFER

To receive a file, you first use a host computer command to display the file on the screen. This requires a knowledge of the operating system, which would usually offer a number of ways to do this. Just before the command is terminated, typically by a carriage return, key in ESC R. After the character R has been entered BSTMS asks for the name of the file to be received. BSTMS then sends the carriage return to terminate the command, and the host computer will begin the file transfer. BSTMS has prepared a memory buffer to store the file as it is being displayed. When the transfer has been completed, which requires the screen to be continuously monitored, the memory contents can be transferred to disk by keying in ESC S. The maximum file length that can be transferred depends on the memory buffer size, and characters will be lost if attempts are made to transfer files longer than the buffer length.

Transmitting a file uses a process similar to receiving a file. Firstly host computer commands are used to open a file for entry from the terminal in interactive mode. Then after keying in ESC T, BSTMS will ask for the file name to be transmitted. At this point BSTMS will start to transfer the file, which can be transferred a line at a time or continuously. There is no limit to the file size that can be transferred as there is when receiving files. A separate feature provided with the BSTMS package is a means to convert binary files to ASCII equivalent, and to recover the original files from the translated version. This only enables binary files to be stored on the host, as they cannot be used there unless a conversion program is written for the host.

As can be seen from this description the file transfer mode in BSTMS is quite limited. The user is limited to single file transfers with size restrictions and there is no error control. The user must also be familiar with the host operating system, which may be unfamiliar territory to the microcomputer user. To achieve more powerful file transfer features, compatible software must be written for both microcomputer and host. This takes the user away from the general-purpose package and into the products designed for individual host computers. Leading the field in this market are the IBM software houses.

Downloading mainframe data to a microcomputer can involve

some tortuous steps if existing systems utilities are to be used. An example will illustrate this: data from an IBM VSAM file is to be transferred to a microcomputer running dBASE II. The mainframe also runs IBM's timesharing option (TSO). The first step is to create a TSO file containing the fields of interest. This is an applications programming job, and a new program is needed for every new query, which can be a source of considerable frustration. The dBASE II user needs to define a file in his system which matches the structure of the TSO file. The TSO file is then down-. loaded by logging onto the mainframe using an asynchronous terminal emulator package with file transfer facilities. The user can then disconnect from the mainframe, rename the file to dBASE II conventions and then hopefully read the file using dBASE II USE and APPEND commands.

If mainframe data is required for insertion into documents, a simpler approach can be used by running a report generator such as FILETAB to produce a printable file containing the fields of interest. This can be downloaded and edited on the microcomputer without concern over the file structure.

As can be seen from this description, transfer of data from mainframe to microcomputer is far from straightforward. The communications software that can provide this missing link must handle the queries, format the files and carry out the downloading with very simple commands from the user.

One of the Informatics General products will be examined to illustrate the power of integrated communications software. Two programs are required to complete the link, dBASE/ANSWER for the IBM PC and Answer/DB for the mainframe. With dBASE/ANSWER running on the microcomputer the user is isolated from the host operating system commands and data base manipulation. These commands still have to be installed in a terminal dialogue file, but this is usually left to the dp staff. Once installed it is always available to the user. The software therefore converts the microcomputer into a smart terminal, and removes the need to get experience with the host operating system.

To acquire the required data, you use dBASE/ANSWER to list the data base names and fields available. After selecting the data base, a series of prompts takes you through the stages of building a data extraction task. When the data extraction task is complete, dBASE/ANSWER establishes a connection between the microcomputer and the host and transfers the task file to the host.

At this stage the host part of the package, Answer/DB, is called into play. The data extraction task file contains all the commands required by Answer/DB to search the required file and format the extracted data. The host software also validates the users' passwords for access to the system and the file. The host search can be run immediately or in batch mode at the dp manager's discretion. Either way, the output is a file which can be retrieved by the microcomputer, using another of the dBASE/ANSWER options. After transmission, the file must be converted to the dBASE format for local data base processing.

The host computer software handles requests from any number of users, and is the key software for other microcomputer applications. Communication between the two programs can be either 3270-type synchronous or asynchronous dial up. The microcomputer software configures the serial interface from simple screen menus. One of the key claims by the supplier is the sophisticated security built into the software. This is particularly important when the microcomputer user is given access to update host data bases.

There is considerable scope for the development of this type of integrated software, and the suppliers are keen to point out their investment in new products. The surprisingly large volumes of personal computers being shipped to major manufacturers gives a clear indication that distributed data bases will be a key market in management and commercial operations.

CASE STUDIES

At the time of writing, the widest experience in connecting diverse microcomputers together and to host computers was to be found in university computing centres. This resulted from the large number of different machines in use for research, teaching and development purposes and in the pioneering role of many universities. Two universities were visited, The Edinburgh Regional Computing Centre (ERCC) and the Bristol University Computing Centre.

A university computing centre provides a wider range of services than is usually found in commercial DP centres. These services may include:

— management services;

— data preparation;

— consultancy – often now on a commercial basis;

— advice to teaching and research staff;

— training for permanent staff and students;

— teaching laboratories;

— software development;

— network communications development with other universities;

— software support, particularly for scientific applications.

Academic freedom also means that users are free to choose a machine on the basis that it is best suited to a particular job, rather than that it conforms to recommended systems. These factors have ensured that there are many diverse types of equipment in a university environment running many different applications. The availability of local hardware and software specialists also means that there are usually resources available to develop customised links between machines. Whilst this is useful in exposing problems and finding new solutions, there has been little output of interest in the commercial market. There have been changes in university structures recently however, and many university-based industrial centres are now springing up, which are purely commercial ventures to top up the reduced funding of many universities.

ERCC is a large centre based on ICL 2980 and dual 2972 mainframes running under the locally developed Edinburgh Multi-Access System (EMAS). These systems together handle up to 200 simultaneous users around the community over a network which gives access to some 600 terminals. The links from various departments are concentrated in three main sites using interlinked packet switching exchanges based on the Joint Network Team design (JANET). Individual mini and microcomputers are con-

FILE TRANSFER

nected to the network via several different types of packet assembly/disassembly units.

Within the ERCC organisation, the microcomputer support unit provides a number of services for the microcomputer user. One of the principal services is the provision of communications software for supported systems. This allows individual users to gain access to the network services, and to exchange files between a large number of different machines.

The strategy used by the Microcomputer Support Group has been to develop a transportable package written in PASCAL for the UCSD operating system. The software can then be moved to any machine which can run the UCSD operating system with a minimum of patching. To illustrate the ease of transfer of the software, known as X-talk, it is currently installed on the following machines:

— Apple II   Kay Pro;
— Superbrain   Sirius;
— Terak   BBC micro;
— IBM PC   Sage 2 and 4;
— Apricot   PDP 11;
— DEC Rainbow   Cromenco.

The communications software was designed to meet the following needs:

— electronic mail;
— secure back-up storage;
— shared access to printers, plotters and other output devices;
— terminal emulation;
— software library.

These needs can be met by a software package that provides intelligent terminal emulation with a file transfer facility for text and binary files. The file transfer function was made more powerful by providing each target host computer with compatible file transfer utilities. It is this step which makes X-talk more powerful than commercial software which uses existing host utilities. It also means that X-talk is difficult to market commercially, although it is freely available to other universities, where development resources are usually available. X-talk has a relatively simple menu display as the machine specific parts are patched in, and the options offered are geared to ERCC services. Transferring the software to new machines and putting in any required patches was said to be relatively trivial, and usually took no more than an hour. Some machines have proved more difficult than others and the Apple II was singled out for criticism as a special assembly language routine was needed to handle the serial interface.

X-talk uses a relatively simple data link protocol which has flow control but no error correction. This was considered acceptable as the transmission quality of the links to departmental machines is very good, and the network connecting the three sites uses X.25 protocols. The host computer software interface ensures that there are no restrictions on file size during file transfers. Binary files were transferred by splitting the 8-bit bytes into hexadecimal equivalents which could be transferred as two ASCII characters. This was also necessary for 8-bit characters which are increasingly used to give extended graphic sets, as some of the network elements could only transmit 7-bit characters.

In addition to the software available on UCSD PASCAL machines, the protocols have been emulated on other machines with a variety of operating systems including:

— BBC micro;
— PC-DOS;
— MS-DOS;
— CP/M;
— C-DOS.

The host computers available to the microcomputer user using X-talk include the following systems, with the language used for the software interface shown in brackets:

— ICL EMAS   (IMP);

FILE TRANSFER 85

— VAX VMS   (IMP);
— UNIX   (C);
— UCSD   (PASCAL);
— Primos   (FORTRAN).

The links between centres are provided by synchronous 2400 baud leased lines using X.25 protocols. A variety of PADs are used to give terminal users access to the network. A number of local area networks have been installed for departmental use. Cambridge rings were historically the first technology used, but for reasons of commercial availability these are being replaced by ethernet type products. The local area networks are provided with gateways and bridges to access the network and to provide communications between local area networks. X-ON/X-OFF flow control is used for the asynchronous terminals and no specific problems were reported other than the 7- and 8-bit data format restrictions. Any potential problems with interpretation were avoided by the use of the 'Hexer' utility to convert 8-bit data and control characters to printable characters only.

Bristol University Computer Centre has a smaller network and is centred around a single site. The central host computer is a Honeywell Multics system with a GEC 4000 as a front end providing X.25 links to other departmental hosts.

A total of 600 terminals are connected to the network in a variety of ways. The access methods used include hard-wired connections to specific hosts, links to a variety of PADs, links to a data switch designed in-house and an increasing use of local area networks.

Bristol has not developed a standard file transfer package, but has had considerable success in introducing the BBC microcomputer as a low-cost terminal. Some 350 machines are in use of which 150 to 200 are used as terminals. The emulator used is a locally developed Tektronix 4010 emulation known as EPIC. In addition to the facilities offered as a graphics terminal, the software has an asynchronous file transfer capability. The development of file transfer software has been held up in the expectation of an asynchronous file transfer protocol standard being introduced. In the interim period a number of solutions have been used to provide file transfer facilities for the wide range of hardware found on the campus. It was claimed that machines could be linked in almost all combinations, but notable exceptions had occurred. One instance recorded was the problem of linking an early Commodore Pet system with the IEEE interface.

The most widely used microcomputers were the following:

—	BBC Micro	EPIC (4010), VT52 and VT100 emulations;
—	Apple	Visiterm, TTY emulation and file transfer;
—	DEC Rainbow	VT52 (supplied), POLY-XFR file transfer software;
—	Cifer 2683	XFILES file transfer via disk;
—	IBM PC	standard asynchronous TTY emulation and file transfer;
—	Superbrain	standard asynchronous TTY emulation with file transfer;
—	CP/M systems	BSTMS TTY emulation and file transfer.

The commercial packages provided simple file transfer facilities via the host editor, which places restrictions on the user. The Cifer 2683 system had the unusual features of a disk-formatting utility which allowed files to be transferred by exchanging disks between machines.

The facilities offered were considered to be adequate as the greatest need was for simple terminal emulation and graphics capability. The main communications applications were listed as:

— terminal emulation;

— file transfer;

— electronic mail.

FILE TRANSFER STANDARDS

The issue of standards is a hot potato and is at the heart of

FILE TRANSFER

conflicting interests of users, who want choice of equipment, and suppliers, who want the user to buy only their products. Standards are emerging, but they will probably only be implemented by some suppliers as bolt-on additions to proprietary network products. It is therefore likely that communications using standard protocols must coexist with proprietary products for a long time. File transfer standards are unlikely to escape this problem as several proprietary protocols are bidding to be accepted as de facto standards at the present time. To illustrate this, British Telecom's manufacturing and marketing wing, Merlin, have announced their support of an asynchronous file transfer protocol known as T-link. This protocol was developed by a US company, Microcom, who market it as MNP, Microcom Networking Protocol, for personal computers. Other data link protocols used in file transfer software include Xmodem (also called the Christensen protocol) and Kermit (available on request from Columbia University Center for Computing Activities).

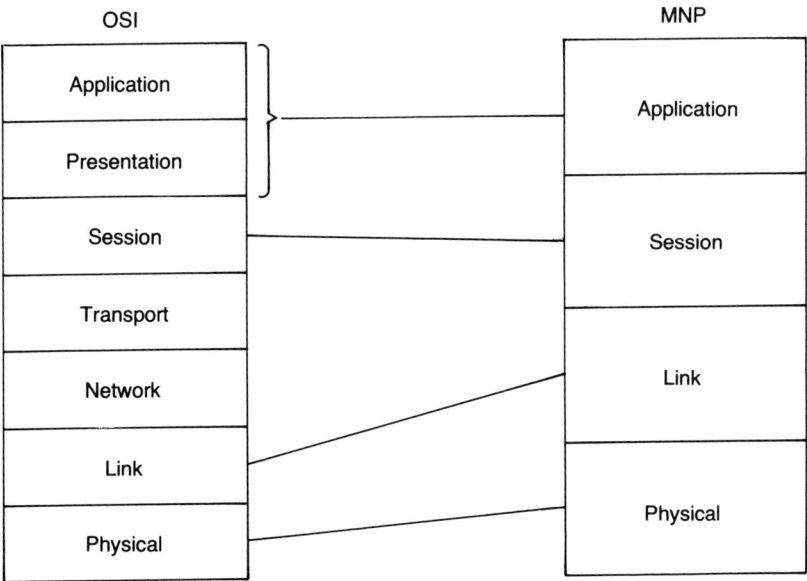

Figure 5.1   OSI to MNP Mapping

File transfer is a high-level service which is located at the highest level, the application layer, of the OSI model. As a high-level service, the services of some or all of the lower levels will be required. Microcom's MNP, for example, is based on a layered concept and can be related to the application, session, link and physical layers of the model, as in Figure 5.1. Although MNP uses the concept of layered protocols, it does not use the detailed standards at all levels as these are not yet ratified. The approach does allow the software to be expanded to incorporate new services as they are required.

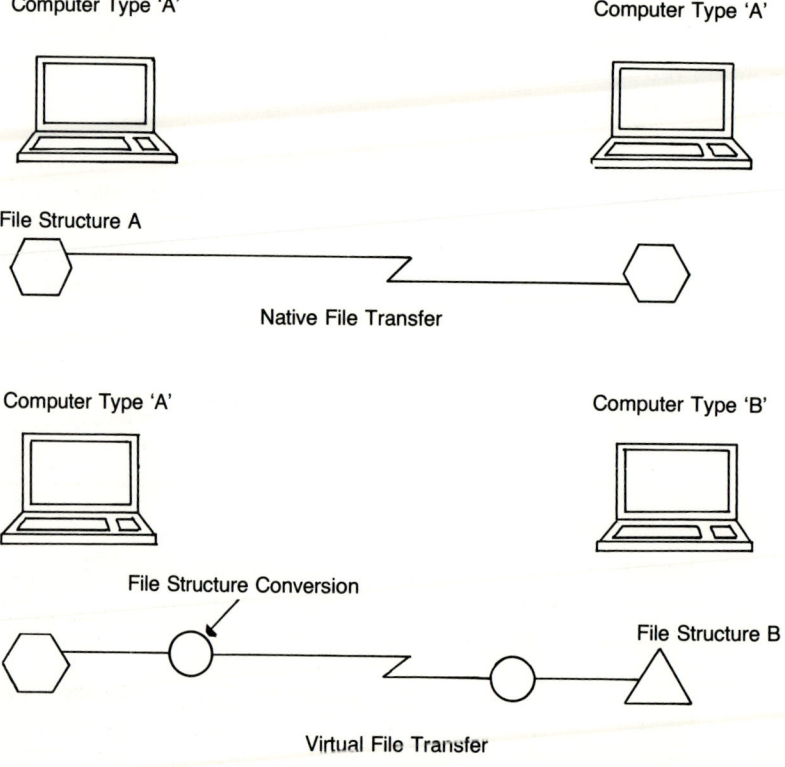

Figure 5.2 Native and Virtual File Transfers

FILE TRANSFER

The four layer protocol used by MNP is designed for point to point file transfers between two machines as shown in Figure 5.2. The file transfers may be between similar machines (native file transfer) or between different machines (virtual file transfer). Virtual file transfer is used to describe the conversion of a native file to a common format which is used to convey the file over the link to the remote machine where it is transformed to the native file structure of that machine. By providing file transformation within the protocol, files with different attributes can be transferred intelligibly, but as was shown previously the application on one machine may depend on a structure which is not available on another machine.

The basis of a layered communications protocol is the data packet, which contains user data enclosed by a number of fields of control information which form the layers of the protocol. This will be illustrated by reference to the Kermit structure which is shown in Figure 5.3. The outermost fields form the lowest level of the protocol (above the physical layer), and the successively higher levels are formed by other fields, each enclosing the fields of the higher levels. At the deepest level is the data for the application layer. The packet is presented to the protocol software by the physical layer, the lowest level, which includes the wire links and interface logic.

As Kermit is an asynchronous protocol, the fields in the packet are constructed from bytes containing ASCII characters. Synchronous protocols are not restricted in this way and bit-oriented protocols are widely used, eg IBM SDLC. The Kermit protocol has six fields to carry out the functions of the three protocol layers (Da Cruz and Catchings, 1984).

The outer fields in the data link layer – Mark, Len and Check – are responsible for the correct detection and demarcation of packets at the receiver, and for checking and acknowledging packet transfers. Retransmission is used to ensure that error free packets are transferred. The session layer – Seq – is responsible for requesting retransmission of missing packets or dumping redundant ones so that the complete file contents are received in the correct order. The application layer – Type and Data fields – contains the status of the packet and part of the file contents which are being transferred.

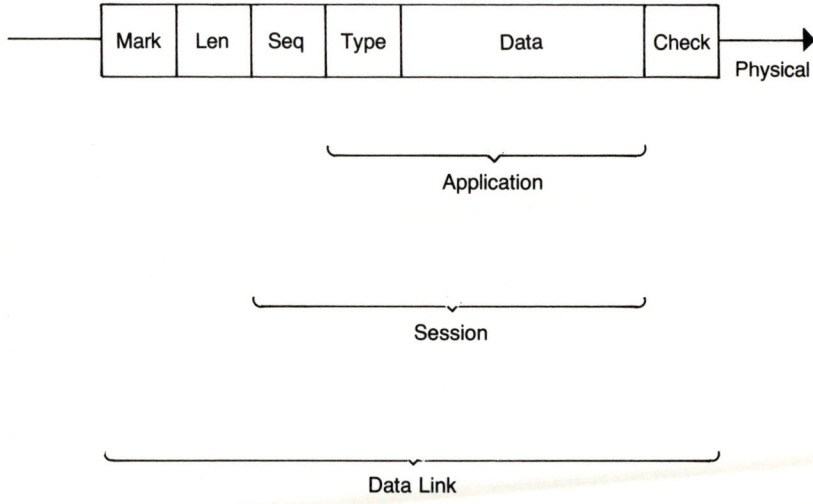

Figure 5.3   Packet Format in Kermit Layered Protocol

Some aspects of the layered protocol cannot be related to the fields in the packet format. The presentation layer for example would be responsible for character conversions (EBCDIC to ASCII) and for the insertion or removal of certain control characters (CR or LF). The Data Link layer also has an effect on the contents of the Data field. This is required for transparent communication, and uses the technique of byte stuffing to modify and flag control characters or 8-bit data words. The technique uses the approach that the simplest way to overcome the difficulties of passing control characters and 8-bit data over the link is to modify all characters to printable characters. The Kermit data link protocol achieves this by prefixing every control character by a special character, normally #, and then complementing the seventh bit (adding or subtracting 64 modulo 64) of the control character. Thus control A becomes the pair of characters #A, and control Z becomes #Z. The prefix character may also occur in the file contents, and must be prefixed itself; ie # becomes ##.

The receiving system requires a data link protocol which can

FILE TRANSFER

perform the reverse transformation. The protocol checks each character and has no effect if the character is any of the printable characters except #. If a # is found, the # and the following character are treated as a pair. If the second character is also a #, then one # is passed to the file. If the second character is any other printable character then the reverse transformation is performed.

Kermit also provides a second prefix for handling 8-bit data when binary files are being transferred. This is used when physical links use the eighth bit for parity and so cannot transfer 8-bit data words. A byte whose eighth bit is set is preceded by another special character, &. If the low order 7-bits coincide with an ASCII control character, the control character prefix # is also added. For example, the byte 10000001 would be transmitted as &#A. The & character can be transmitted by prefixing it (#&). The receiver protocol then has to check for & and # characters and may treat three characters as a single character if the first character is & and the second #.

Prefixing or byte stuffing carries an overhead which depends on the file contents. If the file is a binary file where any data word is equally likely, 50 per cent of characters will be prefixed by &. A further 26.6% will be prefixed by # as they coincide with control characters. A typical text file will have only a small overhead as the majority of characters will be printable.

Another function of the data link layer is data compression, which is implemented in Kermit by a third prefix. The prefix is used when a long string of identical characters is to be transferred. The prefix character ~ is used to indicate that the next three characters are to be treated as a repeat count group. The second character indicates the count and the third character is the actual character in the original file. The repeat count prefix may also be sent as part of the file contents by prefixing it with #.

Byte stuffing or prefixing is used as an alternative technique to hexadecimal conversions for transferring binary files or files with control characters. The hexadecimal conversion (Hex) splits every character into two hex fields, which are then transformed to a pair of printable characters. This is a very simple technique, requiring an elementary conversion process, but it does lead to a constant

100 per cent overhead, which could be costly when dial-up lines are being used.

Error detection is provided in the data link protocol by the Check field. Several options are available in Kermit to give different levels of error detecting sophistication. The simplest base level is a single-character arithmetic checksum. The sum is formed by adding together the ASCII values of each character in the packet except the Mark field and the checksum itself. The resulting sum is truncated to seven bits, which is then transmitted as the checksum character. By repeating this process, the receiver can compare the received checksum with the recompiled checksum, with any difference indicating an error. Clearly some errors will not be detected as the same checksum can be generated by many different character sequences. More powerful block check options are available including a two-character checksum and a three-character cyclic redundancy check (CRC). The CRC sequence is formed from the 16-bit sequence generated by the polynomial $X^{16}+X^{12}+X^{5}+1$ as recommended by CCITT. The high order 4 lists of the CRC go into the first character, the middle 6 into the second character and the final 6 into the third character.

The CRC technique gives a good coverage of various combinations of errors. All single- and double-bit errors, all packets with an odd number of bits in error, all error bursts shorter than 16 bits and more than 99.99 per cent of longer bursts are detected. These figures are taken on the assumption that the Block check has been correctly identified by receiving the length field correctly, and that no intermediate characters have been lost or added.

The function of the six Kermit data fields is summarised as:

Mark   Start of character packet, normally SOH (Control A).

Len   The number of ASCII characters including prefixing characters and the checksum in the remainder of the packet. This is expressed as a single printable character – of which the range 9 to 94 is possible – giving a total packet length of 96 maximum, including the Mark and Len fields.

Seq   The packet sequence number, between 0 and 63.

FILE TRANSFER

	The sequence wraps around to 0 after each group of 64 packets.
Type	A single printable ASCII character giving the packet type.

- D Data
- Y Acknowledge (ACK)
- N Negative acknowledge (NAK)
- S Send initiate (Send-Init)
- R Receive initiate
- B Break transmission (EOT)
- F File Header
- Z End of File (EOF)
- E Error
- G Generic command. The single character following in the data field requests host-independent remote execution of a specified command. Operands may follow the command character.

    For example – L Log out

    – F Finish, but don't log out

    – D Directory enquiry, followed by optional file specification.

- C Host Command. The data field contains a string to be executed as a system-dependent command by the host.
- X Text display header. This indicates that the data field is to be displayed on the screen, for example after a command has been executed at the other end.

Data   The contents of the packet, if any are required for a particular packet type. Non-printable ASCII characters are prefixed with a special character and then converted to printable characters by complementing the seventh bit. 8-bit data can also be sent by prefixing, and a repeat count can be prefixed with a count character. A prefixed sequence of characters cannot be split between characters.

Check   The Block check sequence, based on all the characters between, but not including, the Mark and the check itself. The sequence can be one, two or three characters long. Normally the single-character checksum is used.

The effects of packet corruption can be quite complex depending on which fields are corrupted. For example if the Mark is garbled, the packet will be ignored. Any characters received between packets are treated as noise and are ignored. If any other character within the packet is corrupted into Mark, the current packet will be discarded and a new spurious packet detected.

If the length field is garbled into a smaller numeric value, a character in the data field will be taken as the checksum, and the packet will be detected as corrupt. If the length field is corrupted to a larger value, the second correcting mechanism, the time-out, comes into operation. The receiving station will be attempting to receive data that has not been transmitted, and a software clock is set to interrupt the process if the packet has not been received after a fixed period. Errors which cause times-out are costly in efficiency, but they only account for the small proportion of those errors which affect the Mark and Len fields.

Another function that the protocol provides is file structure conversion. This is another high-level function which affects the contents of the data field and is not part of the outer control fields of the packet. Kermit terminates logical records (lines) in text files with a prefixed carriage return/line feed sequence. File attributes are not transferred, only the file's name and contents, so only text files can be assured of invertible transfers. Non-text file transfers require the user to set up the file structure in the application after the transfer has been completed.

The need to transfer files, particularly spreadsheet files, be-

FILE TRANSFER

tween microcomputers had led to several common file transfer formats being developed. Two formats are in use, fixed and free. Fixed formats define the fields in a record in terms of their position in the row. For example, columns 1 to 10 might be text, and columns 11 to 12 might contain a number. Free formats define the positions of the various fields with delimiters. The most common free format is the comma separated value (CSV) format, in which the comma is the delimiter separating fields, and quote marks enclose text.

Data Interchange Format (DIF) is a proprietary format defined by Software Arts, the Visicalc company, as an ASCII format for interchanging data between applications. DIF was designed to handle spreadsheet data and can describe labels (text) and numbers along with their row and column locations.

To make use of these common file structures, the application program (for example a spreadsheet) must convert the working file to the common structure. This can then be passed to the file transfer software which transfers it to another machine as an ASCII file.

The second machine can then access the file if the spreadsheet application has a utility for handling the file structure used. DIF has become the de facto standard for exchanging data between applications. Most spreadsheets and several data base management products have a DIF file interface built in.

In summary, file transfer products are competing to be accepted as de facto standards, in parallel with the development of international standards. This is the same position that exists in other branches of the computing industry. There is no universal solution yet available, and the complexity of communications protocols illustrates why this is so. At the present time, ad hoc solutions must be used, but a preference should be shown to products that adhere at least in architecture to international standards.

PROTOCOL CONVERTERS

Closely associated with the problem of conflicting standards is the widespread use of matching devices known as protocol converters. As the need for interworking has grown, many companies have introduced protocol converters which allow two different machines to communicate, each in their native protocol. The technique requires two interfaces linked by a data buffer. Data from one device – say, a mainframe with a synchronous protocol – is received by the protocol converter which handles the polling requests. The data is stored in a buffer before being passed to the other device over the second interface. The buffer not only allows the storage of data during the polling sequence, but also allows different line speeds to be used.

Protocol converters are available to match a great many different proprietary protocols. A typical application is to connect an ICL or IBM mainframe using a synchronous port to an asynchronous terminal. The terminal could be a microcomputer running a terminal emulator with a file transfer function. File transfer is possible because the protocol converter handles the polling, leaving the microcomputer free to transfer the file from memory to disk using X-ON/X-OFF flow control. The previously discussed problem that the microcomputer was not powerful enough to respond to I/O requests and transfer memory contents to disk does not arise.

This approach would only be useful if a large investment in microcomputers has already been made, as the cost of a protocol converter, at around £1000, is high compared to the cost of a more powerful microcomputer. The economics become reasonable when a protocol converter with multiple asynchronous ports is used with a single host port.

6 Local Area Networks

THE NEED FOR LOCAL AREA NETWORKS

So far the emphasis has been on direct connections, either between microcomputers or between microcomputers and host computers. Whilst it is unlikely that the need for direct connections of this type will disappear, the growth of distributed computing is likely to dominate computer systems in the future. Ever since computers were first introduced into commercial, manufacturing and research organisations the need to provide links between machines was recognised. Larger organisations began to develop computer networks and found the need to provide various network services, such as routeing, error-free transmission protocols, etc. The computer manufacturers responded to this need by developing proprietory network architectures, such as IBM's system network architecture (SNA). The PTTs also offered public data network services, such as British Telecom's Packet Switch Stream (PSS). These network architectures provide the basis for developing a wide area network, extending across site boundaries right up to a global scale.

Wide area networks developed alongside the growth in the mainframe computer industry. As terminals and new types of digital equipment proliferated a number of problems began to concern the managers supporting this equipment; for example:

— there is often considerable difficulty in connecting machines from different suppliers so that files and other information can be exchanged intelligibly;

— the speed of the existing networking equipment was usually

too slow to meet all needs adequately;
— the conventional wiring systems were expensive to install, and were inflexible and difficult to expand.

These problems prompted the search for new techniques to provide the required connectivity for a wide range of equipment, combined with flexibility and low cost. The prime target for these new developments was the office system, which was recognised as a large, unexploited market. Amongst the equipment appearing in the office can be found word processors, personal microcomputers, minicomputers and facsimile machines. Teletex equipment will soon be added to this growing list of complex, microcomputer-based equipment. Yet it is often impossible to exchange information intelligibly between these machines. It was in this context that local area networks were introduced, promising to solve the many problems of interconnection and interworking.

The techniques used in today's local area network products have their basis in a number of different developments. The computer bus is historically the earliest technique used to interconnect different digital functional units along a common highway, although limited to a single system. This approach was developed by Xerox and others to a bus system for interlinking minicomputers. Alongside this development, computer communications based on packet switching had become a viable technique as demonstrated by the Arpanet. In this network, a common network transmission medium was shared by a number of users by bundling data into groups or packets, which were identified by address, sequence and other labels. It was a logical step to use this technique for buses designed for intercomputer communications over a limited area. At the same time as the bus system was being developed, an alternative approach emerged, possibly inspired by the electrical ring main for domestic electricity supply. The University of Cambridge's Computer Laboratory was the main innovator in the development of a data network based on a ring of interconnected machines. The two approaches, bus or ring, have one common feature, that they provide a single physical medium for interconnecting all the equipment attached to the network. The method used to provide a number of simultaneous connections is the responsibility of the access technique. Although access techniques are quite different for the alternative bus or ring approaches, their

LOCAL AREA NETWORKS

aim is the same, to provide a high-speed link between devices attached to the network with low probability of data loss or error.

There is likely to be some convergence in the distinct types of local area network presently available as the gap between microcomputers and mainframes becomes smaller and one or two local area network techniques become dominant. A range of networks is required to cover the range of applications from the smallest office-based organisation up to the largest industrial and commercial sites. This requirement is a matter of scale and will be reflected in speed of transmission which in turn governs cable, interfacing and repeater costs.

Included in a local area network package would be the physical transmission medium, which might be a twisted pair, coaxial cable, multi-core or fibre optic cable, and cable taps, network interface units and software which allow individual computers and other devices to be attached to the common transmission medium. The essential feature of a local area network is that it can cover all locations contained within one site, up to a size unit which depends on the technology used, and is therefore entirely under the control of one organisation.

Although the physical transmission medium is confined to one site, this does not prevent users of the network communicating with other public or private networks outside the site. Access to services outside the boundaries of the local area network is provided by attaching devices known as bridges and gateways to the network. A bridge is used to describe a link between two similar networks, and a gateway for a link between the local area network and a different type of service such as Telex.

It is not the intention in this book to cover local area technology in detail as this has been fully described by Gee (1982) and Flint (1983). Details of the technology will be limited therefore to that required to show how local area networks can provide interconnections over a common physical medium.

LOCAL AREA NETWORKS
Main Characteristics

The restriction on the range of the network to within the boundaries of one site opens up a number of options to the designer. In particular the structure or topology of the network can take several different forms and even more options are available when choosing the transmission technique. The transmission methods available are vastly different to those commonly used in wide area networks as supplied by the common carrier. This is mainly because the network is confined to one site so that new physical transmission lines can be installed and interference levels can be controlled. High transmission speeds can be achieved (0.1 to 50 Mbps) over coaxial cable or optical fibres with relatively low-cost, line-driving equipment and repeaters. Because of the scale of local area networks (from 1km up to about 10km) and the uniformity of the transmission medium, error rates are usually very low. Local area networks can therefore be summarised as networks confined to small areas, operating at high transmission rates with low error rates, and providing interconnections between a large number of diverse computers and peripherals.

Local area networks have the following characteristics:

— cheap transmission medium;

— cheap devices to interface to the transmission medium;

— easy physical connection between computer-based equipment and the network;

— high data transmission rates;

— the network provides a means for interconnecting equipment operating at different speeds;

— connections can be set up between any devices attached to the network;

— the network usually operates without central control, giving high reliability;

— it is easy to extend the network.

It is important to note that these characteristics are not dependent on the transmission technique, topology or access method used. As the technology has become established, the type of equipment that can be connected to local area networks has

become more clearly defined. The list of equipment includes:
— personal computers;
— computer terminals (dumb and intelligent);
— host computer systems (minicomputers and mainframes);
— mass storage devices and file servers;
— printers and plotters;
— copying equipment;
— process control equipment;
— bridges and gateways to other networks;
— some forms of telephone equipment.

Applications

Although there are many shapes and sizes of local area network, the primary purpose is to provide a communication channel between computer-based equipment. The communication channel may be used to transfer information between people, to provide access to a shared data base or to share expensive equipment. There is a long-term aim for local area network designers to extend the applications to real-time video, analogue voice and colour or black and white picture transmissions. These requirements can be met by the broadband approach which is a development of cable television technology. Details of integrated services, local area networks and broadband technology are discussed by Gee (1982) and Flint (1983).

Local area networks have a number of characteristics which can be used to advantage by the applications programmer. Some of these are:

— *High-speed transfer.* A transmission speed of 10 Mbps is typical for a moderate local area network, and even allowing for access time and overheads an average speed in excess of 1 Mbps can be achieved for a lightly loaded network. As has been previously mentioned, the transmission is not continuous but split into packets, and the effect of loading is to increase the mean time to deliver a packet. The local area network design can be scaled up or down with correspondingly higher or lower transmission speeds to cater for different sized sites. Even the most moderate systems still offer a transmission speed that is an order of magnitude higher than that possible in conventional systems using analogue telephone lines.

— *Rapid establishment of circuits*. Connections made over the telephone system are characterised not only by the limited transmission speed available but also by the slow signalling speed and long call set-up time. In contrast, local area networks can establish a path between two attached devices very quickly, and can maintain the path for the duration of the data transfer. This type of connection is quick, it appears to the user that he is the sole user of the network and the connection is usually referred to as a virtual circuit. The time taken to transfer the initial packets in the circuit establishment phase would typically be less than 10 milliseconds.

— *Only small delays in data transfer*. The performance of a local area network under load when all the attached devices require network service is of major interest to prospective customers. Fortunately the network operating speed is so much greater than the average data rate for all users combined that overloading, which appears as excessive delay, rarely becomes a problem. The different technologies do behave differently under load, either having a predictable delay (deterministic networks) or a randomly varying delay (contention networks).

— *Speed matching*. The facility for speed matching is inherent in the concept of local area networks. The use of a common high speed transmission path and an access technique based around data buffering makes this possible. The external interface is usually a V.24 serial interface, which can be operated at any of the usual baud rates, but other interfaces can be provided. The serial interface can also be operated synchronously or asynchronously, although ring architectures are more suited to synchronous operation than bus architectures, as will be shown later. Asynchronous ports can be operated with the normal choice of character formats. The technology does then promise great flexibility in providing connections between different equipment, but in practice the hardware design of the network interfaces usually means restrictions have to be observed.

— *Resource sharing.* A key economic factor in favour of local area networks is the ease with which expensive equipment can be attached to the network to be shared by all users. Perhaps the most common example is the shared single printer. This can either be connected to the network via a server which buffers data for printing, or can be connected directly and individual users are queued for service by the network. If several printers are available a further service which can be offered is hunting for a free printer and then establishing the circuit. The technique of queuing can be extended to computer ports which are usually connected to terminals via a multiplexer or front-end processor in a conventional network. In a local area network one computer port can be shared by several terminals using a queuing technique. If the port is busy when a terminal tries to make a connection, the network monitors the current connection and automatically releases the circuit and connects the second user when the first transaction is completed.

The other main resource that is shared is the file store and this is also managed by a server. This has to keep track of all the accesses being made and correctly assemble file data into packets for transfer to the user's device. It is often the performance of the file server which is the critical factor in the overall network performance.

— *Security.* To prevent users from connecting to every port on the network, individual ports can be assigned passwords to restrict access to those users with the correct authority. Another type of security service is the provision of closed user groups. A closed user group consists of a group of ports which are so designed that messages can only be transferred between them, and information cannot pass into or out of the group.

— *Broadcasting and conferencing*. Broadcasting is a facility for transmitting a message from one terminal to all others on the network. They may be sent out immediately or at some prearranged time. Some restrictions may have to be placed on broadcasting, particularly for synchronous terminals which can only receive and transmit data using the correct polling sequence. Suppression of broadcasts would also normally be offered as an option. Conference calls allow a group of users to send messages to all members of the group and to receive messages from any member of the group. This facility does pose problems of management, to prevent a screen-based conference becoming garbled due to messages overlapping.

— *Transparency*. Local area networks should be able to accept and convey all possible bit patterns. In practice the degree of transparency offered is variable, some suppliers claim that the data field can be used for any protocol whilst the others place restrictions on the protocols supported. Before a transparent circuit is set up, a dialogue is required to configure the network or network interface unit for the required protocol.

— *Gateways and bridges*. These are terms used to describe the equipment used to provide links between networks. There is some disagreement over the precise definition of gateways and bridges, and sometimes other terms, such as filter, are found. The distinction made here is to use the term *bridge* when referring to a piece of equipment linking two similar local area networks, and a *gateway* for equipment linking networks with quite different protocols. More precisely, the bridge describes a link between networks which use the same network layer protocol. A gateway has to provide an interface between higher protocol layers and is correspondingly more complex.

— *Low error rate*. Low error rates are inherent in the design of local area networks. In one trial (Shoch and Hupp, 1979) a detected packet error rate of 1 in 2×10^6, or one packet in error per day, was achieved. As a data link protocol which incorporates retransmission is used, detected errors are not

LOCAL AREA NETWORKS

passed to the application. The undetected packet error would be better than 1 in 10^{11} which is adequate for most process control applications and more than adequate for text transfer. These low error rates are several orders of magnitude better than can be achieved on the analogue telephone network.

In addition to low error rates, local area networks also have a high reliability. Reliability is strongly influenced by design factors such as whether a central control is required and how the network responds to cable breaks and access point failures. A target for total downtime of less than 0.02 per cent has been proposed by Flint (1983).

— *Remote access.* A primary aim of local area networks is to enable terminals or computers in one location to work successfully with another system in another location as if they were all local to each other. This implies that performance is uniform across the network and that intervening modes are transparent to the user.

The list of features and facilities just described enables the network designer to provide applications that would be difficult or expensive with conventional networks. The conventional data circuit suffers from a number of failings: it is slow to establish, difficult to manage, has a high error rate and operates only at low speeds. A number of new applications will be described together with the ways these can be implemented using local area networks. A new term which has recently emerged is the workstation, which is used to describe a personal computer system customised for the office environment and possibly incorporating a telephone handset.

In addition, there are facilities for:

— *Electronic Mail.* Electronic mail systems allow users to transfer text messages easily and quickly. To provide a comprehensive electronic mail service a number of facilities are required:
  — a text editor for writing messages;
  — a mechanism for sending messages to one person or a group of people;

— storage for messages sent and received;
— a mechanism for answering messages which have been received;
— confirmation of delivery;
— privacy for confidential messages;
— a priority system for sending urgent messages.

Each user needs to have a workstation available and all the workstations must be interconnected. Although a completely distributed electronic mail system is possible, there are advantages in having a single message system device which incorporates routing control and message filing.

The local area network is an ideal basis for an electronic mail system as it has the capacity and scale to enable all users to be interconnected with the required reliability, speed and security. New users can be added as the need arises and the flexibility of the network allows equipment to be moved with minimum disruption of service.

— *Distributed Database*. There is a wide variety of information to which an individual in a large organisation requires access. This information ranges from strictly personal data such as diary details, names, addresses and telephone numbers up to corporate data held on mainframe computers.

The workstation linked to a local area network provides an ideal means for managing such a distributed database. This is usually achieved by having a central file store which is partitioned into individual user areas and common user areas. In addition local back-up storage is usually provided for each workstation based on floppy disk technology. The local area network has the speed, reliability and flexibility to provide users with the services they require from various types of data file.

Access to external data bases such as Prestel can also be provided to network users via gateways. With a choice of gateways, more than one external service is available from a single workstation. Before local area networks became

LOCAL AREA NETWORKS 107

available it was usual to have to use a variety of different terminals to access these different systems.

There are other traditional applications and services which can benefit from implementation on local area networks. These include:

— *Terminal concentrators.* The bulk of terminals still in use are dumb devices which need intelligent devices to control them. Where a group of terminals is located on one site, a local area network can be used to provide this control and concentration function. The ease of installing new terminal access points and the availability of the local area network cable running throughout a site makes the dp manager's job very much easier. The local area network solution would usually be reserved for situations where a large number of terminals are located on one site. Where a number of small sites are involved, traditional methods would prove to be more economic.

— *Resource sharing.* This is often the prime reason for any organisation investing in a local area network. The equipment which is usually provided as a shared resource includes:

— large capacity hard-disk systems;

— line printers;

— letter quality printers;

— digital plotters;

— gateways and bridges to other networks.

— *File transfer.* As a local area network provides a reliable, fast interconnection service which can be extended to all locations over an entire site, some of the problems of file transfer have been solved. The electronic mail and distributed data base applications require a basic file transfer service, which can be developed to give a more general facility. Not all local area networks offer full 8-bit data transparency, so a protocol which transfers files as printable characters may still be needed. The use of data buffering and packet transfers can also result in problems in certain circumstances. These problems will be investigated in a later section.

Technology

The restriction of the local area network to one site opens up a lot of options to the designer, and a wide variety of different techniques have been developed into successful products. The main requirements which were described in the last section can be summarised as:

— low cost;
— high transmission speed;
— sufficient capacity to meet projected load;
— low error rate;
— high reliability;
— easy to connect to;
— able to take mixed types of traffic;
— able to be linked to other networks.

The design of a local area network can be broken down into four aspects, which overlap to some extent:

— physical transmission medium used;
— network topology;
— transmission mode;
— access method.

The first three areas are closely related and should be discussed as a group. Access method is a major category and has greater significance when comparing local area network products than the other areas.

Physical Transmission Medium

The choice of physical transmission medium is quite wide and includes coaxial cable, twisted-pair cable, flat-ribbon cable, fibre

LOCAL AREA NETWORKS 109

optics, radio and infra-red transmission. The selection process is governed by cost, speed and interference factors, but there is some justification in going for a high-quality medium as it should be a one-off purchase, and replacing an inadequate system would be very costly in both financial terms and disruption caused. Against this argument is the development of better and cheaper techniques, particularly in fibre optics, which may well become the system of choice in the future. At the present time interfacing to fibre optics is rather more difficult than to other media, and fibre optics are not compatible with some other design features.

Network Topology

The topology of a network describes the way the nodes are interlinked. Four topologies can be identified which can meet the needs of a local area network. These are the star, the loop, the ring and the bus.

A star network (Figure 6.1) has a central control system with connections radiating out like the spokes of a wheel. The most widely known star network is the telephone system which is based on a number of interconnected stars. The mainframe computer system is another field where star networks are widely used. It is usual to build a star network as a cluster of small stars for reasons of economy.

A loop network (Figure 6.2) consists of a central controlling device connected to the two ends of a cable loop. All the terminals and other devices are connected to the loop and use it to send messages to other devices on the loop. The essential feature of the loop is the single controlling device which controls individual access to the loop. The loop topology was developed as a terminal concentrator system, using low speed transmission lines.

A ring network (Figure 6.3) is a development of the loop in which a set of nodes are connected together in sequence so that each node is connected only to the nodes either side of it. The nodes have equal status and take no part in controlling the network. A central monitor is required to maintain the operation of the network, and to initialise the network. A central control is not strictly required, but the problems of errors building up and faults developing make the use of a central control device desirable.

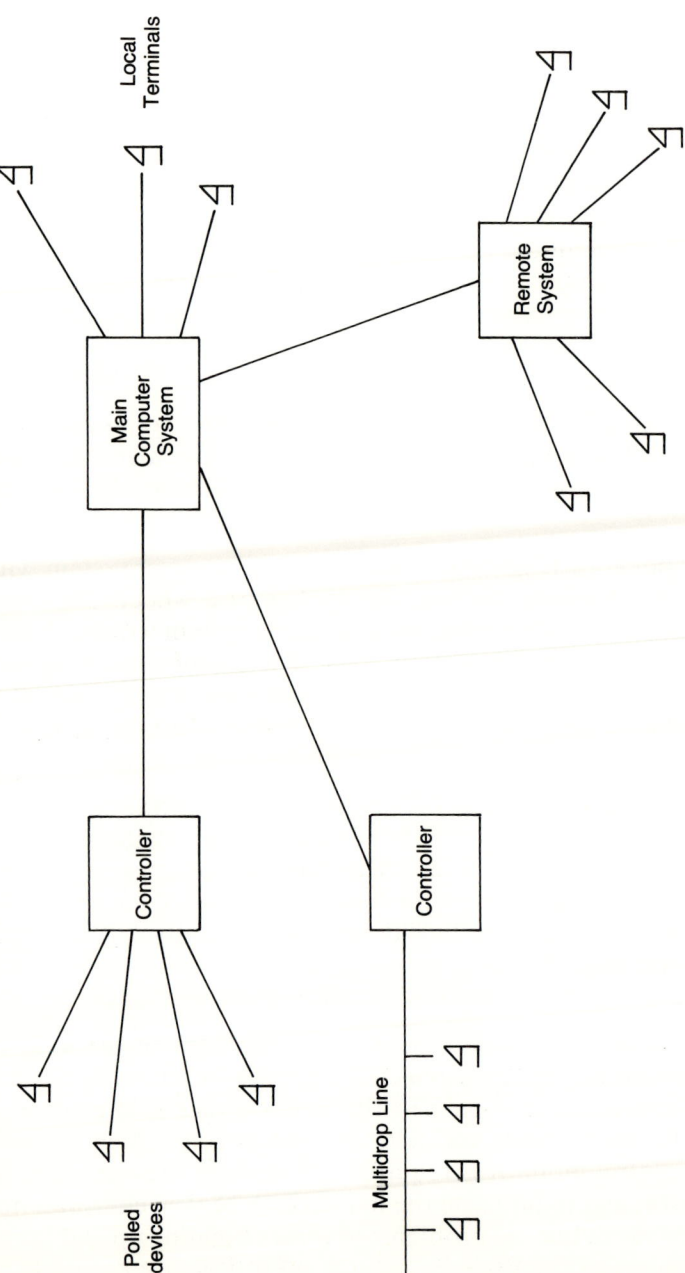

Figure 6.1 Star Mainframe Computer System

LOCAL AREA NETWORKS

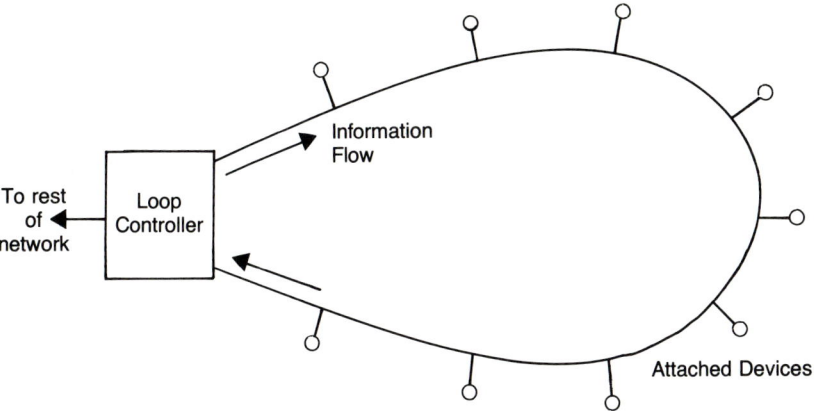

Figure 6.2 Loop Network

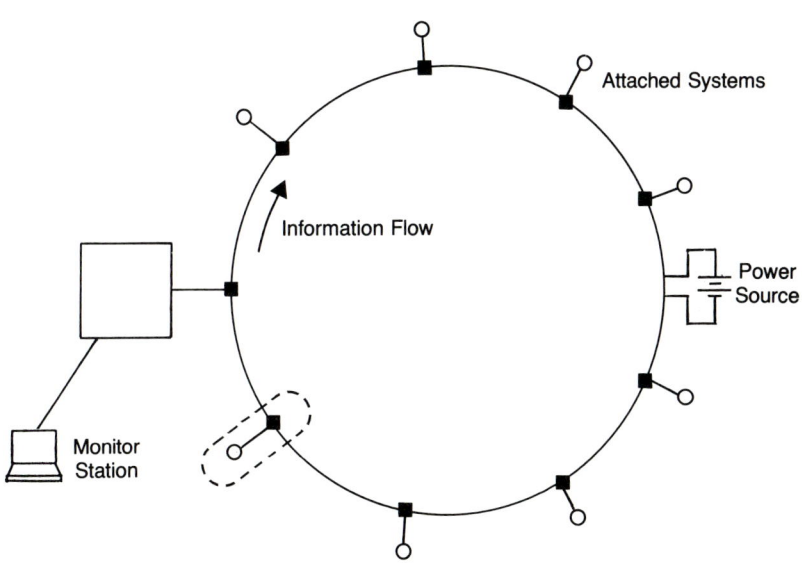

Figure 6.3 Ring Network

112 MICROCOMPUTER COMMUNICATIONS

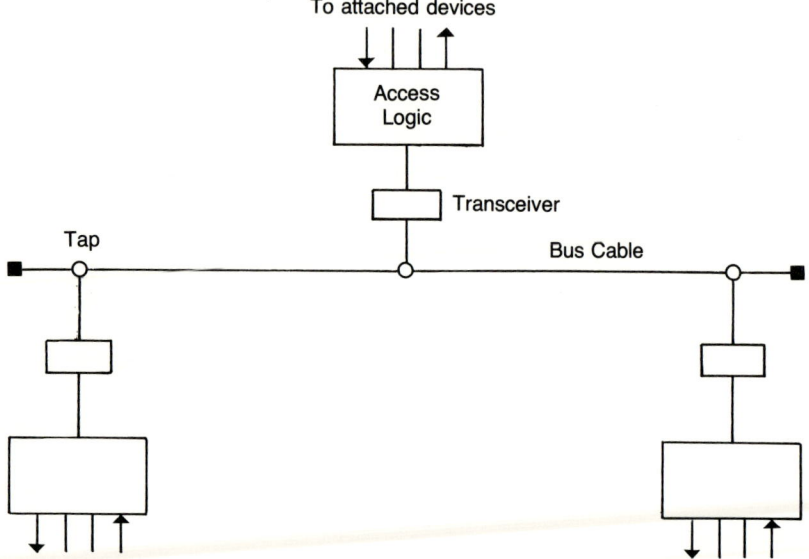

Figure 6.4 Bus Network

A bus network (Figure 6.4) consists of a highway, which may have branches, providing a single route between all devices connected to it. It is a development of the computer data bus, which is built into computer systems for interconnecting the main functional units such as central processor unit, memory and peripheral controllers. The major difference between bus and ring structures is that the bus is completely passive and does not need to be broken to add new nodes. This is possible by using a type of coaxial cable which can be tapped by piercing the outer sheath with an insulated probe. This technique cannot be applied to fibre optic cables however which are difficult to integrate with a bus structure.

Transmission Mode

The transmission mode describes the kind of signalling technique used to carry the data from one location to another over the physical medium. For local area networks two classes of signalling technique are used: Baseband and Broadband.

LOCAL AREA NETWORKS

Baseband signalling, which is the simplest method in use, employs discrete changes in signal level to represent the binary information content of the transmitted data. Of the various ways of encoding the signals, Manchester encoding is the most widely accepted. A sample data stream which has been converted to Manchester encoding is shown in Figure 6.5. It has one feature which makes it very valuable in the design of communications systems: it carries clocking information with the data stream which simplifies synchronisation design. Each time interval is divided into two, with the signal level in the first half representing the complement of the binary value and the signal level in the second half representing the uncomplemented value. In this way there is always a transition midway through the bit interval.

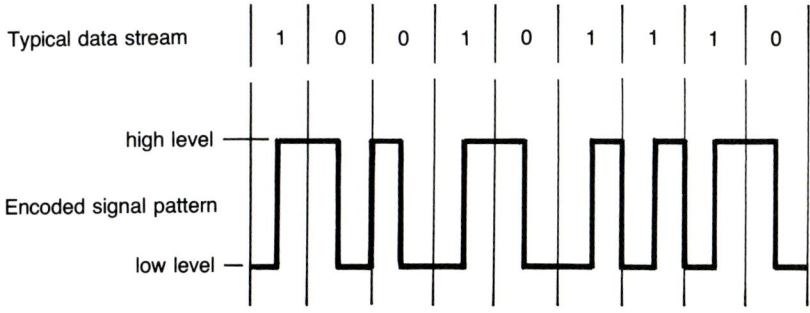

Figure 6.5   Manchester Encoding

In *broadband* signalling, the digital data stream modulates a high frequency carrier signal which is continuously present on the transmission medium. This technique is well established for cable television (CATV – Community Antenna Television). The bandwidth of CATV coaxial cable is around 300 MHz, and there is ample bandwidth for data transmission services and other services such as real time video. A single frequency slot can be used to supply the same services as a baseband system.

The signalling method used depends on the physical medium

used. Coaxial cable can be used for baseband and broadband signalling, whereas fibre optic cable is only suitable for baseband as the signal can only be modulated in light intensity.

The signals applied to the physical medium are subject to attenuation, distortion and noise whether baseband or broadband signalling is used. For long cable spans analogue amplifiers for broadband systems or digital repeaters for baseband systems are used to provide error-free transmission. The distances to be covered also dictate the type of physical medium to be used: for example, twisted pairs for 50 metres, coaxial cable for up to 500 metres and optical fibre for up to 5000 metres.

The transmission properties of the physical medium are not the final limitation to network size as it has been shown that amplifiers and repeaters can be used for long-distance, error-free transmission. The fundamental restriction on network size is the result of the last class of technological choice, the access technique.

Access Method

It is the choice of access method which is the most hotly debated issue of the present time. The access method overlaps with network topology as access methods can dictate the network topology to be used. The access method is closely linked with the way the network operates and is the most fundamental aspect of network design in any comparison of different systems. It has been left to this stage to show how equipment is interfaced to the network and how different systems compare in the flexibility and simplicity of the connection method.

Access methods can be divided into four broad classes:

— selection or polling techniques;

— contention techniques;

— reservation techniques;

— other techniques appropriate to the ring topology.

The first three methods can be used in the ring topology, but the special characteristics of the ring allow access techniques to be used which cannot be transferred to other topologies. Access

LOCAL AREA NETWORKS 115

methods are discussed in detail by Gee (1982) and Flint (1983) and will be briefly summarised in this section.

Polling. Polling has already been referred to in connection with synchronous transmission. In a polled network each station is allowed to transmit when it has received permission from a central controller. This is the technique which will be familiar to mainframe users where a front-end processor controls a number of terminals connected to a multi-drop line.

Polling techniques are well developed and chips are available to allow implementations to be economically designed. The technique is fair to all attached equipment and the response time can be accurately predicted.

On the negative side, a polled network is vulnerable to failure in the central polling unit and overheads are high and access time relatively long. The technique has been successfully applied to several small-scale local area network products, but is not considered suitable for large-scale systems.

Contention. Contention is the technique usually associated with the bus topology. Contention schemes allow attached devices to use the network at any time and provide a means to solve the problem of simultaneous attempts to use the network. The attractions of the technique are that for lightly loaded networks access time can be very small, the network is resilient, and reliability is good as the bus is passive and there is no central controller. In addition physical interfacing can be very simple if coaxial cable is used and devices can be attached or removed very easily. The main disadvantage arises when the bus becomes heavily loaded as this leads to an increasing number of clashes between competing devices and delays in transferring data become significant. Studies have shown that packet delays vary exponentially with load, which would make contention buses unsuitable for some process control applications.

There are a number of contention schemes which have been successfully used:

— Carrier Sense Multiple Access (CSMA);

- Persistent CSMA;
- Carrier Sense Multiple Access with Collision Detection (CSMA/CD);
- Carrier Sense Multiple Access with Collision Avoidance (CSMA/CA).

These long descriptions are indicative of the complexity of the techniques. Carrier sense means that before a device attempts to access the bus it first listens to it to find out if another transmission is already taking place. Multiple access means that the transmission medium is shared by a number of users. Persistent is used to describe a listening technique which allows the device to begin transmission as soon as the end of the previous transmission is detected, or with some probability P. Collision detection is a mechanism which allows devices to recognise that another device has started to transmit at the same time and a clash has occurred, requiring both devices to back off. Collision avoidance is a combination of fixed time slots and CSMA/CD which gives devices a fixed time slot under certain conditions.

The choice of contention technique affects the throughput that can be achieved on a particular channel. The contention techniques are linked in order of complexity and throughput varies from less than 20 per cent of the channel capacity to over 90 per cent for the most sophisticated schemes (Flint, 1983). The most widely implemented scheme for local area networks is CSMA/CD, which is the scheme used in Ethernet.

Contention is a simple concept and allows the use of a passive cable which need not be broken to add new devices. The main disadvantage is the statistical nature of the response time. A second major disadvantage is the need to fix a minimum packet size and network size to enable the collision detection method to work.

Reservation Techniques. Reservation schemes divide up the capacity of the network by allocating time slots to the attached devices. Time slots can be allocated in a fixed manner using a standard time division multiplexing technique but there are other more efficient systems:

- static allocation, set up during system configuration;

LOCAL AREA NETWORKS 117

- call allocation, set up for the duration of a call;
- dynamic allocation;
- hybrid call and dynamic allocation.

There are other reservation techniques particularly for ring topologies, but because of their special characteristics they are described separately. Reservation techniques have advantages when different types of traffic are mixed but there is a higher cost overhead compared with other systems.

Ring Access Methods. Ring networks have some special characteristics that allow access methods to be designed to exploit the symmetry of the ring. One particularly well known technique is the slotted ring, in which a number of fixed length data packets travel round the ring, passing each node in turn. This is the technique used in the Cambridge Ring.

An alternative approach to allocating a fixed group of rotating packets is to allow nodes to insert packets when a free space is found. This technique – register or buffer insertion – requires the nodes to be based on repeaters which can switch a shift register buffer in or out of the ring circuit.

Another widely used scheme is known as token passing, which requires a special packet, or token, to be circulating around the ring. A node must wait for the token to be received before it places its own data packet on the ring. Token passing is claimed to be efficient and fair, and in common with other ring systems the response time is deterministic.

Ring systems are all based on active transmission paths and almost always require a central monitor to maintain the slots or token, which can give rise to concern over reliability.

Connecting to a Local Area Network

With such variety in technology, a few examples will be given to show how equipment is connected to representative products.

Ethernet

The Ethernet specifications are available from the Xerox Corporation, and can be used by a manufacturer for a nominal licence fee. This has led to Ethernet being adopted as a standard within the ISO reference model. The specifications have now been incorporated into the IEEE 802.3 local area network specifications. A typical implementation is shown in Figure 6.6 which shows the physical units and their relationship to the overall architecture. The Ethernet specification (Digital, Intel and Xerox, 1980) covers the physical and data link layers of the OSI model, which form the basis for higher level protocols to be added by individual implementors.

The physical interface consists of a tap which can connect directly to coaxial cable by piercing the outer sheath to make contact with the inner core. This is the feature which makes the system attractive to the network manager. The cable is a 50 ohm type and must be designed so that the pressure tap does not introduce excessive losses or distortion. The connection is then buffered by the transceiver unit before the access logic and device interface complete the connection of the equipment to the bus. The illustration shows the access logic and device interface implemented on a board which would be installed in the equipment. This controller board is then connected to the transceiver using a cable which carries send data, receive data, collision detect and power in four twisted pairs with an overall shield.

As Ethernet is a specification rather than a single product some variation in implementation can be expected. One early implementation, Net One from Ungermann-Bass, was designed as a terminal network, and offered network interface units with serial V.24 ports in various configurations. This design predated any large-scale integration and had to group ports together to obtain reasonable economy. The Ethernet controller and transceiver installed inside equipment is an economic proposition now as chips have become available.

The data link layer, which is provided by the controller, receives data packets from the attached device and builds frames by adding type, address and error checking codes. Complete frames are then passed from the controller to the transceiver which broadcasts them to all other devices attached to the bus. The specification allows 10^{14} possible addresses which could easily be arranged to

LOCAL AREA NETWORKS

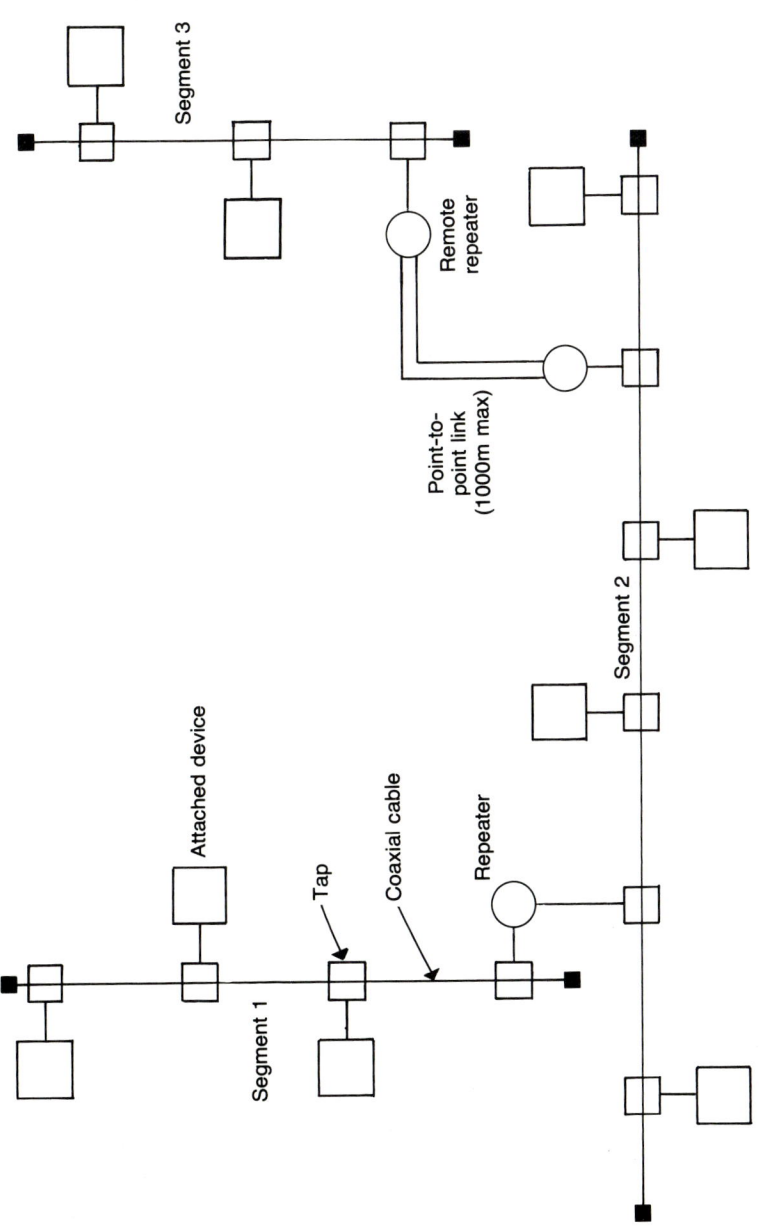

Figure 6.6 Typical Large-Scale Ethernet Configuration

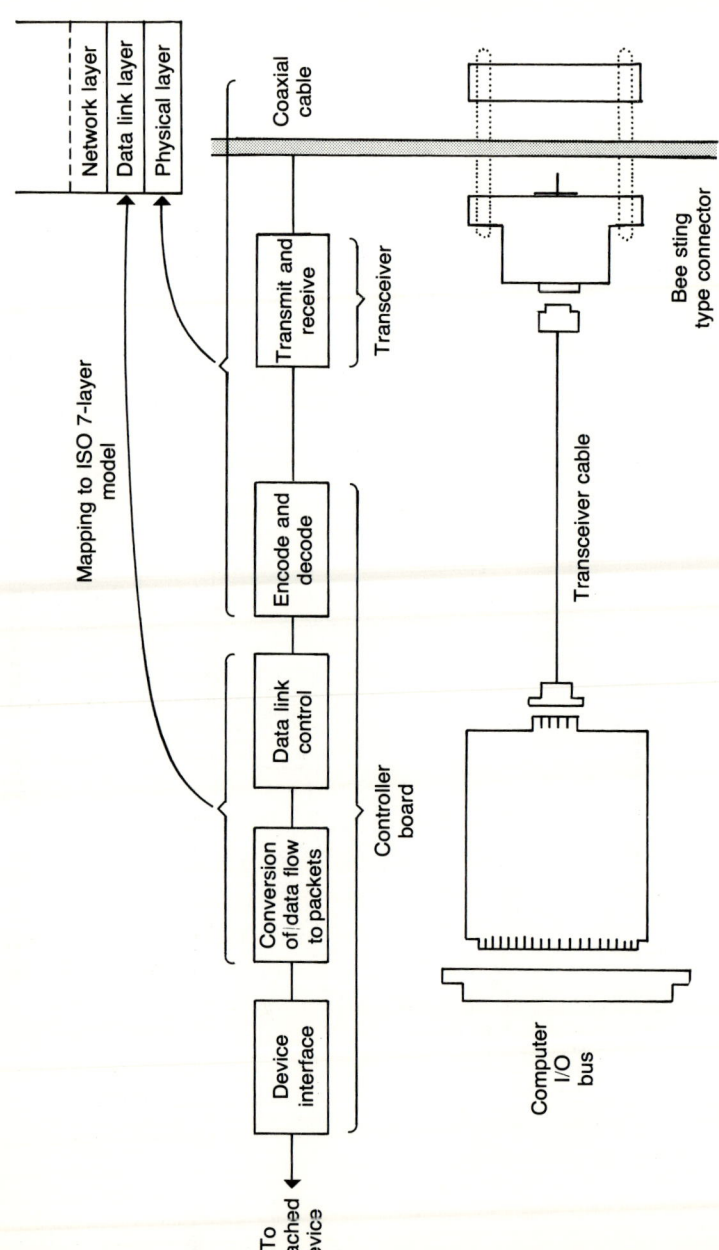

Figure 6.7 Typical Ethernet Implementation and Functional Organisation

LOCAL AREA NETWORKS 121

give world-wide coverage.

Ethernet buses can be linked together by repeaters to give the network designer flexibility in the network layout. A typical site layout is shown in Figure 6.7. The access technique places restrictions on the transmission quality of the network, and rules must be adhered to in order to guarantee satisfactory performance. The main limitations are on cable length, and the avoidance of significant reflections which could result in misoperation of the collision detection circuits.

Planet

Planet (Private Local Area NETwork) from Racal-Milgo is a ring topology network using the token passing access method. It is intended to offer a series of products aimed at the information processing market.

Planet uses a standard twin coaxial cable, of 75 ohm impedance, which is widely available from a number of suppliers. The ring topology and token passing access method do not require the same high cable standard as a contention bus system.

Access to the ring is by cable access points (CAPs) which are passive devices, and small enough in size to fit into or on cable ducting. Connection to the cable is by BNC connectors at each CAP. A typical network is shown in Figure 6.8. The philosophy is to place CAPs at regular intervals wherever there is a present or potential requirement. CAPs which are not in use sit passively on the network and do not interfere with the operation of the active CAPs.

Terminal Access Points (TAPs) are connected to CAPs and provide the interface for attached equipment. TAPs are responsible for regenerating the ring signals and passing packets on to the next TAP. The TAP includes all the higher level protocols and each TAP has two V.24 ports for attaching terminals and other equipment. Up to 250 TAPs can be accommodated on each ring giving a maximum of 500 connections. This can be increased by using multiple rings.

In addition to the individual CAPs, TAPs and their attached devices, a central monitor, the Director, is also attached to the ring. This retains a copy of the system configuration on EPROM

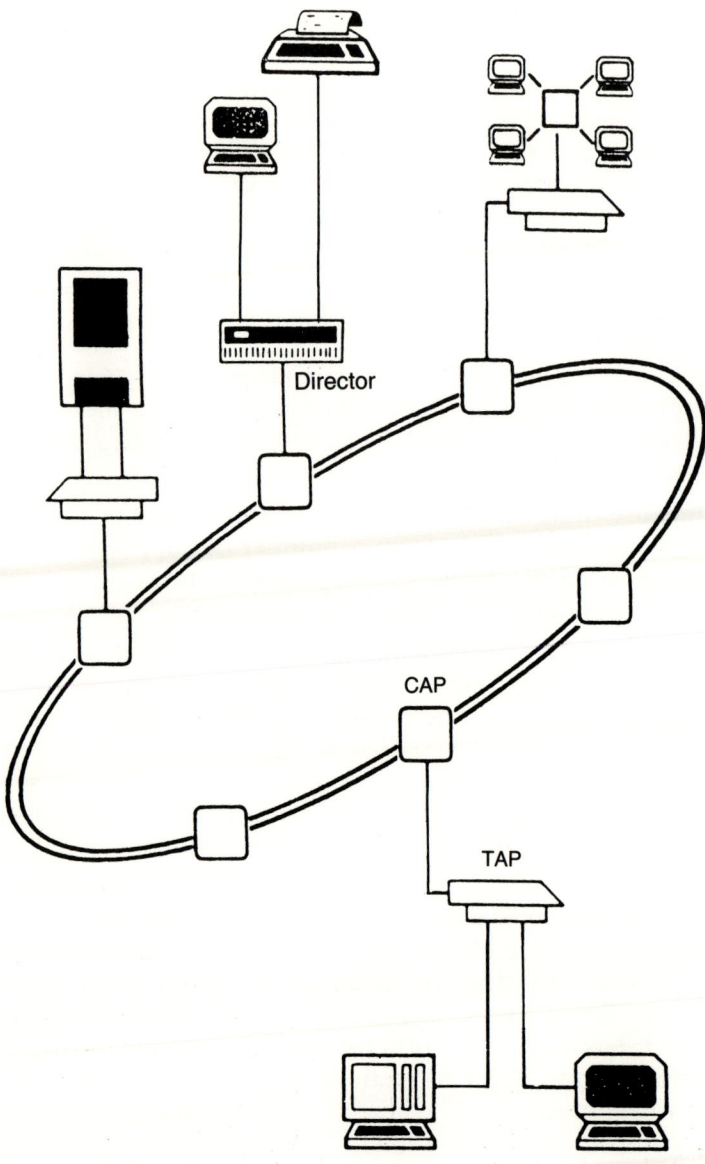

Figure 6.8 Planet System
(Reproduced by kind permission of Racal-Milgo Limited)

LOCAL AREA NETWORKS 123

(CMOS RAM) which is reloaded whenever the network needs to be rebuilt. The Director stores details of port names and TAP addresses together with associated peripheral information. It also performs a network management role over all aspects of ring operation including call set up, transfers and new configurations. All packets pass through the Director, and if an error condition is found in a packet it is removed and normal operation of the ring restored.

The folded ring topology is used to give increased reliability by introducing redundancy into the network. Cable breaks or component failures can be eliminated by reconfiguring the network as shown in Figure 6.9. As can be seen the ring is looped back at either side of the Director so that although the CAPs form a completed circle, data packets do not travel right round the ring.

The Director also provides statistical data which includes a number of parameters:

Call statistics:
— number of attempted connections;
— number of successful connections;
— number of unsuccessful connections;
— total connection time.

Network statistics:
— number of ring breaks;
— number of parity errors.

The type of services provided indicate that the network was primarily designed as a terminal network. In this role a variety of virtual circuits can be set up. The connection patterns are known as plans, which are stored in the Director, and include the following possibilities:

— permanent virtual circuits;
— switched plan virtual circuits;
— switched virtual circuits;
— call queuing and hunt groups;

MICROCOMPUTER COMMUNICATIONS

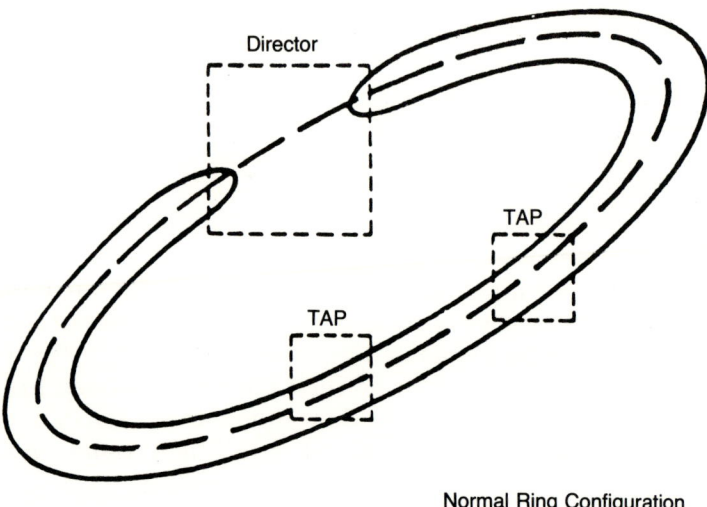

Normal Ring Configuration

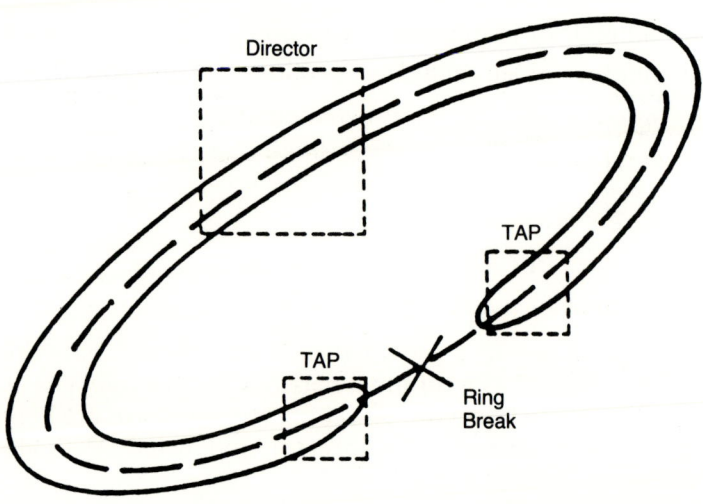

System Reconfiguration

Figure 6.9 Planet Resilience Mechanism
(Reproduced by kind permission of Racal-Milgo Limited)

LOCAL AREA NETWORKS

— conference switched virtual circuits.

Permanent virtual circuits are long-term connections between any ports on the ring. Permanent virtual circuits can be direct port-to-port connections, multi-drop connections or conference calls. A multi-drop circuit consists of a master port which broadcasts to a number of slave ports. Each slave port can only transmit to the master port.

Switched plan virtual circuits offer the same types of connections as in permanent virtual circuits, but the configuration can be changed from any authorised terminal. Both permanent virtual circuits and switched plan virtual circuits are preserved over power failures.

Switched virtual circuits are short-term connections between two ports, which are set up by the user of an asynchronous terminal. These circuits are cleared on power failure, and shut down by the Director or when the user disconnects. The circuits can be set up by one of the participants in the call or by a third party.

Call queuing is a feature available to callers using switched virtual circuits that allows a request to set up a circuit to be queued until the destination port is free. A second type of queuing circuit is allowed, the hunt group. When a connection request is made to a member of a hunt group the connection will be made if the port is free. Otherwise a connection will be made to the first available free port in the group.

Conference circuits can be provided as permanent or switched circuits and offer some special features. A conference circuit allows a group of terminals to be connected together so that they can all transmit to each other and receive each other's transmissions. This facility raises an interesting question of management as it is possible for messages from two or more users to be garbled if they try to transmit at the same time.

Gateways and Bridges

In order to meet the needs of users of local area networks wishing to extend communications outside the network, gateways and bridges can be provided as ports on the network. The concept of the gateway or bridge was introduced earlier in showing how different types of link are required between networks. Gateways have been examined in detail by Gee (1983), and the results will be summarised here.

Gateways are pieces of equipment used to connect local area networks to the long distance networks such as the public telephone system, packet or circuit switched data networks and private networks – collectively known as wide area networks. The gateway is more precisely called a LAN/WAN gateway to distinguish it from other types of gateway as the term is a generic name for any device which links dissimilar systems.

Local area networks and wide area networks have such different characteristics and protocols that gateways are of necessity complex devices. Because of this, and also due to the relative newness of local area networks, gateways are expensive and few suppliers can be found. There is general acceptance that these devices will be needed but the requirements are not yet fully understood. Most suppliers have indicated their intention to market gateways as soon as possible, particularly for connection to the telephone network and packet switched services. Initial products are likely to be dedicated to specific inter-networking functions as a full-function gateway is liable to be too expensive for most customers at the present time. By no means all of the problems in gateway design have been solved and there is a considerable amount of research work being undertaken.

Experiences with gateways so far have been disappointing. They have not been able to insulate the user from the differences between the two networks and are therefore far from user-friendly. This is because early samples do not have the higher levels of software built in which could take over the long-winded setting up operations from the user. Also as single user devices, there is the problem of queuing for a valuable service. It is necessary to install a number of gateways in a hunt group to avoid the frustration of finding the single gateway busy with no alternative. Another problem which has been encountered is the slow speed of some gateways. Not only are set up times long, but the time taken to process data reduces the throughput possible so the effective transmission speed is reduced.

The problems of gateway design are not insurmountable, but there is still some way to go. The requirements are for gateways which combine ease of use with a wide range of facilities and at a low cost.

7 Wide Area Networks

INTRODUCTION

Computers and their associated peripherals can be linked together using leased lines or switched circuits provided by the PTT to form a wide area network. A wide area network can range from just a few sites separated by distances measured in miles or fractions of a mile up to the global networks of major multinationals.

Wide area networks have until very recently been built around analogue telephone circuits, and the data transmission has been characterised by low speeds (typically 1200 or 2400 baud) and high error rates. Conveying digital data over an analogue line, band limited to 300 to 3400 Hz, requires modulation using modems operating at a variety of speeds and prices. A rapid change is taking place in the mid-1980s as the telephone network is converted to an all-digital system. This will result in a common transmission technique for all types of traffic and will offer higher data rates with much improved error rates.

The provision of a physical network of transmission lines is not sufficient to provide a complete network service. To make any two pieces of computing equipment exchange data intelligibly requires monitoring and organisation. This requirement is met by a set of rules or protocols which must be coordinated, the end result being a network architecture.

A network architecture is a framework for interconnecting computers and their related peripherals in such a way that they can exchange data regardless of their different functions and internal design.

As a network is extended outside a single site the available transmission techniques are restricted to those supplied by the PTT, and this results in wide area and local area networks having quite fundamental differences. One of the most obvious differences is in the error rates of the transmission links used. The higher error rates found in leased lines or switched lines means that error handling procedures must be incorporated in the wide area network basic link-level protocol. Another factor is that a network covering a number of sites over a wide geographical spread cannot use a symmetrical topology or bus design as found in local area networks. The important factors in determining how the individual sites are linked will be their geographical locations, estimates of traffic between sites and economic factors. The resulting topology will usually grow as a series of stars connected together by a mesh of interconnecting links.

The differences between wide area networks and local area networks are not limited to differences in the physical link and topology as an even greater variety of access method can be found in the wide area network. Although network design based on packet switching protocols may appear to be a natural choice, circuit switching networks should not be ignored. Modern PABX design in a digital transmission environment means that fast circuit set-up lines can be achieved with low error rates and so circuit switching can be a viable alternative, particularly as line costs will be reduced as digital circuits become available. It is really too early to say just what the impact of the Integrated Services Digital Network (ISDN) will be on the future development of computer networks.

It might appear from the previous discussion that switching, whether it is circuit or packet switching, is an essential element in wide area networks. In fact relatively few networks require switching facilities, and the majority of commercial networks are built around point-to-point links. The real benefits in choosing a packet switched architecture are in the network management facilities offered in the package.

As there are so many different options open to the network designer it is not surprising that the problems that can arise for the microcomputer user linked to a wide area network are quite

WIDE AREA NETWORKS 131

diverse. This will be examined by first looking at the broad classes of wide area network in use, in terms of the elements used in their construction. Networks can be divided into the following main groups:

Private networks
- based on proprietary network architectures;
- based on public switched services;
- based on standard network architectures;
- custom designed.

Public data networks
- Telex;
- Telecom Gold;
- Other Value Added Networks.

PROPRIETARY NETWORK ARCHITECTURES

The best known proprietary network architecture is IBM's System Network Architecture (SNA). Although this is not a standard system, IBM have decided that it is in their interest to publish the protocols, and other companies have been able to design products for the SNA market. The market position of IBM is so strong that SNA has achieved a dominant position, whilst OSI protocols are still some years away from completion.

SNA is a comprehensive design, based on packet switching, that provides support for the complete range of IBM products. One way for the microcomputer user to interface to an SNA network is to run a 3270 emulation which is possible for many of the microcomputers in use. Although the first IBM PCs did not have a 3270 emulation mode, this became available with add-in cards, the most popular being the IRMA board from Digital Communications Associates Inc. IBM soon followed up the original PC product with a range of microcomputers, including the 3270 PC which has a full 3270 emulation mode as standard. Whilst these products allow access to an SNA network, communications software has not been announced by IBM to extend the applications open to the user from simple terminal transactions to true distributed processing. The user is therefore still dependent on the independent software houses for the vital communications software.

This approach can equally well be applied to other proprietary network architectures including Digital's DECnet, Honeywell's DSA, ICL's IPA, Burrough's BNA or Hewlett Packard's DSN.

The main advantage in choosing a proprietary architecture and single supplier is that the user can have reasonable confidence that the complete family of products can interwork satisfactorily and an upgrade path will be offered. On the negative side other suppliers may have better and cheaper solutions to some requirements, but they cannot be connected to the network.

PUBLIC DATA SWITCHING NETWORKS

A private network can also be built around a public service, either circuit switched or packet switched. Because of the deficiencies in the public switched telephone network for data transmission, public circuit switching is of little importance at the present time. To cater for the network designer wishing to use a public service, British Telecom has introduced the Packet SwitchStream (PSS) service. This has the dual attraction of low cost and high reliability, and has a growing user group. The cost benefit arises from the tariff structure which relates cost to the volume of data sent. This can be favourable for some users who might otherwise have used leased lines for either low data rates or bursty transactions.

As PSS is a public service, the access method for a microcomputer user will be examined in some detail. PSS was developed from an earlier experimental service using protocols which did not fully agree with CCITT recommendations in the X series. PSS uses CCITT recommendations and there are now a number of X-series products which can be used to interface to the PSS network. The main X-series recommendations are:

    X.25   This specifies the packet interface to the network. The protocol provides an error free transmission link over the network.

    X.3   This specifies the parameters used in the packet assembler/disassembler (PAD) to host communications link, and therefore the facilities implemented in a standard PAD.

WIDE AREA NETWORKS   133

X.28   This specifies some details of the PAD to terminal interface, including PAD commands and service signals for a standard PAD.

X.29   This defines the protocol for non-data messages passing between the PAD and the host computer. These messages allow the host to read and/or set the PAD parameters, which therefore gives control over the host to terminal dialogue.

THE PACKET ASSEMBLER/DISASSEMBLER

The most important piece of hardware is the PAD which provides an access method for terminals and computers which only have a V.24 serial interface. PSS users have the choice of installing a PAD on their own premises or using a BT PAD located in the Packet Switching Exchange. In the second case the PAD is accessed via a dial. The PAD converts the serial data stream into X.25 packets which can then be handled by the packet switching exchanges. Packets are unpacked at their destination by a remote PAD or host computer. The relationship between the various X-series recommendations in an asynchronous terminal to host computer link are shown in Figure 7.1.

The rest of this discussion is about PAD operation using one of the best known products, the CAMTEC JNT-PAD, for illustration. This was developed by CAMTEC in collaboration with the

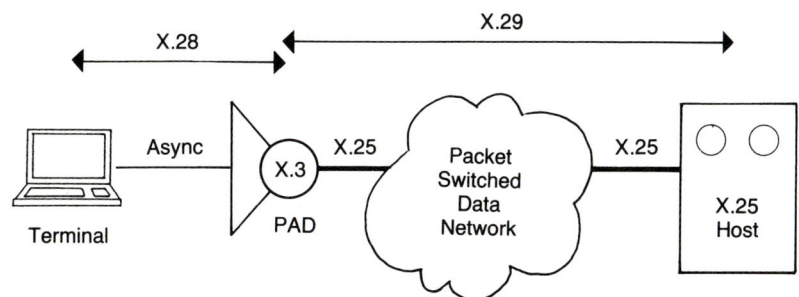

Figure 7.1   Relationship between the Triple X Recommendations
(Reproduced by kind permission of Camtec Electronics Limited)

Computer Laboratory of the University of Cambridge and the Joint Network Team of the University and Research Council Computer Centres. Although originally designed to the Joint Network Team (JNT) 'Rainbow' book standards, the choice of X.25 at the packet interface level has led to the PAD having applications outside the academic network, in particular, British Telecom's PSS service.

The PAD offers the following facilities:

— up to 16 asynchronous terminals can be connected to a single PAD, with switching and routeing to X.25 networks on up to 10 lines;

— in addition to the assembly and disassembly of X.25 packets, input lines may be edited and output formatted;

— calls to host computers connected to an X.25 network may be made, controlled and cleared;

— a packet switching exchange function is available by switching incoming X.25 calls to become outgoing calls on the same or any other X.25 port;

— the JNT 'Yellow Book' Transport service is supported to provide a routeing service within one network;

— terminals connected to the PAD can make calls to other terminals on the same PAD, without involving X.25;

— 8 opto-isolated 20 mA current loop ports can be provided in place of the V.24 interface card;

— an option provides autoload and start on power-up with site specific configuration data stored in memory;

— where a user normally connects to only one host, there is an Autocall facility which automatically sets up a call to the pre-selected host when the user presses the return key;

— where a user requires to connect to a series of hosts, a PAD command allows calls to be set up using a short form address;

— if a call fails, the PAD can set up the call via an alternative route;

— to provide assistance to a new user, a Help command is available which sends details of commands, terminal profiles, addresses and escape sequences to the user's terminal.

The PAD is designed to be transparent to the user as far as possible. Thus the networking services which provide, for example, re-routing of calls are invisible to the user. Once the call has been set up, the PAD enters the data transfer state in which the terminal appears to be directly connected to the host. There are some important differences from a direct connection however, including the optional use of a local flow control protocol between the PAD and the terminal, and the necessity of holding data in buffers between the terminal and host interfaces.

The user can return to the PAD command state at any time during a session by pressing the break key or control P and A. A null command (ie Return) will cause the PAD to revert back to the data transfer state. The data transfer state operates in one of three modes: Message, Native or Transparent.

In Message mode, the exchanges between the terminal and host are line-at-a-time, and the PAD is responsible for the correct presentation of this data on the terminal. The PAD maintains a 'terminal description' of the terminal hardware and this is referenced during PAD to terminal transfer. The PAD echoes all characters typed at the terminal keyboard and provides local line editing and output formatting.

In Native mode the PAD passes characters received from the terminal to the host without processing. Input is forwarded to the host a character at a time at normal typing speeds. The host is responsible for echoing input and the editing functions. The PAD acts transparently to all 8-bit input and output with the exception of Break and Control P/A, which invoke the PAD command state.

The Transparent mode allows the host to take over control of output formatting so that terminals other than simple scroll mode devices can be attached to the network. This mode is suitable for graphics terminals.

Normally the user is unaware of the mode the PAD is operating in, as this will either have been fixed at installation time, or will change during the transaction under the control of the host by manipulation of the X.3 parameters.

Another feature which simplifies the operation of the PAD is the Profile command. Profiles are stored parameter lists which contain the terminal description values. Profiles are configurable by the site manager and are easily selected by the user. This feature is particularly useful when one port is used by a number of users with different terminals, for example using dial-up access. The terminal description describes the terminal hardware and includes such items as line width, parity, terminal type (VDU or printer), tab handling and flow control. Some of these parameters are not recommended as yet in X.28, but are included to give the widest possible cover of different terminals. The VDU/printer parameter, for example, changes the PAD responses in the editing mode. If VDU mode is set, a BS SP BS sequence is sent in response to the delete character, 'erasing' the deleted character from the screen. If Printer mode is set however, a "/" character is echoed after the delete character to indicate on the printer that one character has been deleted.

A second set of parameters is maintained by the PAD for interactions with the host. These deal with host parity, tab handling, echo, line feed, insertion, etc.

When the host sends Clear or Reset PAD X.29 messages to the PAD, English text messages are sent to the terminal to help the user interpret and diagnose the fault condition that had occurred. The messages can be changed on-site.

A PAD is a sophisticated network device, and is able to provide a number of services with either simple, easy to use commands or with no user interaction at all. This is possible because of the extensive software built into the PAD. A few examples of PAD applications will illustrate the versatility of the PAD in a variety of different networks.

Figure 7.2 shows the most common application for the PAD, as a means of connecting asynchronous ASCII terminals to X.25 networks. The terminals can be asynchronous VDUs, printers with keyboards, printers, plotters or videotext terminals (ie Prestel). Dial-up access over the PSTN can be provided by connecting

WIDE AREA NETWORKS   137

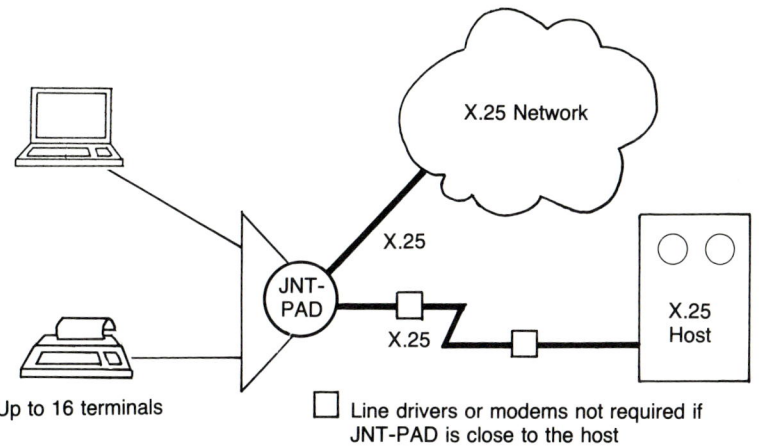

Figure 7.2 Normal PAD Operation
(Reproduced by kind permission of Camtec Electronics Limited)

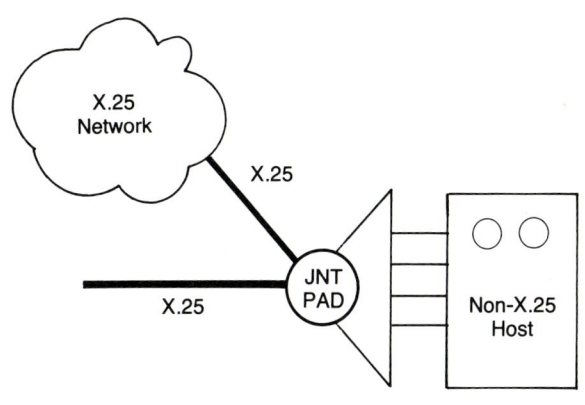

Figure 7.3 Reverse PAD Operation
(Reproduced by kind permission of Camtec Electronics Limited)

autoanswer modems to the V.24 ports. As the CAMTEC PAD has two X.25 ports as standard, access to two X.25 networks can be provided. The network shown consists of one port connected to an X.25 network with the second port connected to a local X.25 host. In this configuration the terminals can access the wide area network and the local host, and in addition the local host can access the wide area network using the switching capability of the PAD.

Figure 7.3 shows the PAD connected in an alternative reverse mode. In this example a host with a number (up to 16) of V.24 ports is connected to the PAD. The PAD offers a 'hunt group' facility in this mode and will connect an incoming call to the next free port of a predefined hunt group. Calls to specific host port addresses can also be made.

Figure 7.4 shows an extensive system, including normal PAD operation, reverse PAD mode and an extended number of X.25 ports (available with a switch option). This shows the versatility of the PAD in meeting a very wide variety of data communications needs.

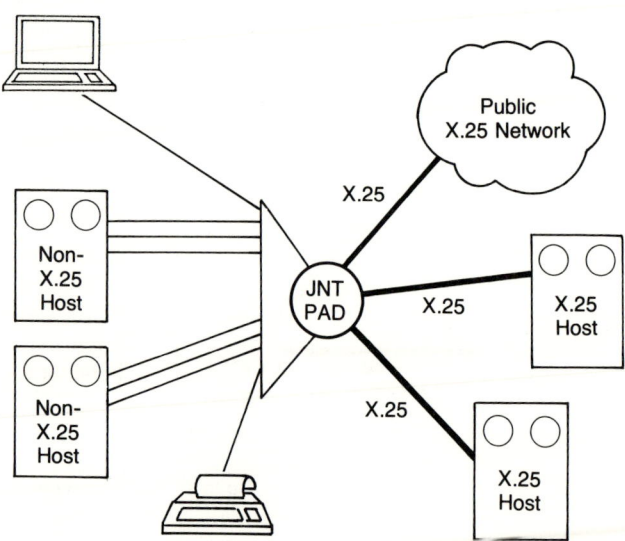

Figure 7.4 Large Network with a Combination of PAD Modes
(Reproduced by kind permission of Camtec Electronics Limited)

WIDE AREA NETWORKS 139

Figure 7.5 shows the use of PADs as an alternative to statistical multiplexers. The diagram shows two ways of using the PAD as a concentrator. Often a leased line will be the most economic solution, but the British Telecom PSS service can be used as an alternative. This is a particularly beneficial solution when links to more than one site are required.

The flexibility and power of the PAD concept make it a very valuable element in the design of complex networks. The PAD is not limited to wide area networks, and the majority of University computing facilities are based around X.25 local area networks using the CAMTEC JNT-PAD. These local area networks are then linked together to form the JANET wide area network for the academic community.

Whilst the X.25 and associated recommendations provide powerful connection services it must be remembered that these are low level services in a fully distributed computing system, and do not provide, for example, interworking and file transfer. It is also important to note that, even in transparent mode, the PAD cannot transfer all 8-bit characters because of the need to provide a control sequence to break to PAD command level. This is charac-

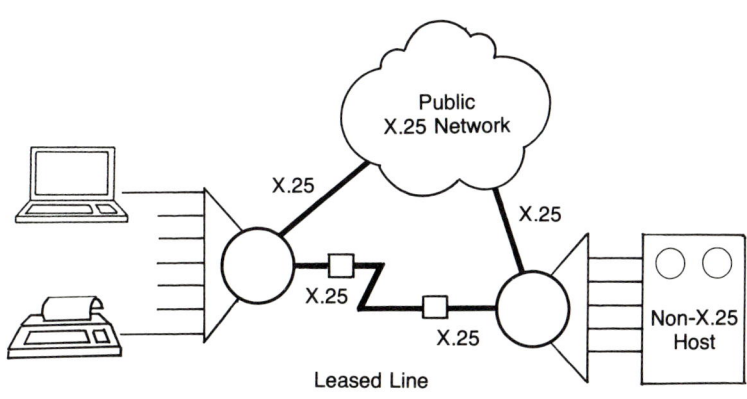

Figure 7.5 Use of a PAD as a Line-Sharing Device
(Reproduced by kind permission of Camtec Electronics Limited)

teristic of many intelligent network devices. File transfer applications will therefore still require a low level protocol which transfers files as printable characters.

THE DATA SWITCH

Circuit switching has already been mentioned as a viable alternative to packet switching for some network requirements. Data switches based on circuit switching principles can either form the backbone of a computer network, or sometimes be a component part of a network which uses packet switching on the main links between sites. The choice is made on cost and functional performance. The main selling point for data switches is the high level of interconnection that can be provided at an economic cost. These points will be illustrated by reference to the Gandalf PACX which is the best known data switch at the present time.

The PACX system is highly modular and gives the network designer great flexibility in designing an economic network, with provision for expansion. The basic PACX switch offers the following facilities:

— system capacity in terminal connections: 256 up to 1024 in steps of 4, with a minimum installation of 4;

— system capacity in host port connections: 128 to 512 in steps of 4, with a minimum of 8;

— synchronous and asynchronous transmission supported;

— wide range of data rates: maximum for asynchronous terminals 9600 baud and for synchronous 19200 (varies with total system capacity);

— terminal queuing provided;

— up to 128 terminal profiles may be stored;

— security is provided at two levels: access to hosts can be barred and a password feature can be selected;

— monitoring and reconfiguration are provided on-line;

— a log of all connections and disconnections is available on a statistics port;

- fault diagnosis is simplified by the use of a monitor which can tap into the data flow through any port;
- different service classes can be defined;
- direct terminal-to-terminal connections can be set up;
- PACX to PACX switching is available to build up more extensive networks;
- optional interfaces: RS 232 C/V.24/V.28 as standard, RS 422, RS 423, RS 449 and 20 mA current loop as options.

This wide range of facilities and options allows the PACX switch to be used in a wide variety of network configurations. The main application is to connect terminals to host computer ports, giving maximum access to the terminal users. Digital switching techniques are used to give fast set up times with contention and queuing services built in. In addition network management facilities are built in to monitor the switch performance and to streamline the switch operation.

The data switch approach has some advantages over networks based on packet switching concepts. As data buffering is not required, and a permanent circuit is held between the terminal and host port, 7- or 8-bit data transparency can be offered and problems of delay and control code interception do not occur. Conversely, the system cannot provide speed matching or data concentration over a single line. Synchronous transmission can also be supported with few potential problems.

A number of compatible options are supplied by Gandalf to extend the application of the PACX switch. These include gateways to X.25 networks and protocol convertors for the IBM 3270 series.

PRIVATE NETWORKS – CONCLUSIONS

It is only possible in this book to introduce the subject of wide area networks. The previous sections have shown the wide choice of network design and components available to the designer. The choice of technology or service used depends on many factors including type of business, geographical spread, existing hardware, applications and personal preference. There are few hard and fast rules in network design. There is one factor, however, which may change the design process radically and that is the emergence of international communications standards. Implementation of OSI standards holds out the promise of full interworking between different vendors' equipment by using a common set of communications protocols from the lowest (physical) layer to the highest (application) layer. If OSI is successful then all proprietary and public network architectures will have to move towards the OSI standards.

Even when the OSI standards have become firmly established there is little sign that the present mixture of local area networks, data switches, integrated data and voice PABXs and wide area networks will change. Each approach suits a particular environment, but the common feature could be all these different technologies working to OSI standards.

CONNECTION TO PUBLIC DATA SERVICES

Some of the basic components of a data network have been described in the previous sections. In this section some of the important public data services will be described, with some guidelines on how the microcomputer user can access them.

Telex is an important service for one reason, it is the only international data service at present in use. The service is quite limited in that only text can be transferred between Telex terminals, and the speed, 50 baud, is very slow. The character set is also very restricted comprising the upper-case alphabet, the numerals 0-9, 13 other characters and the space.

The service uses old technology working asynchronously, and data transfers frequently result in lost or corrupted characters as the transmission quality is poor and error correction is not provided. Nevertheless, the Telex service is one of the services which a business microcomputer user is likely to want to access. Telex is the subject of international agreements, and has over one million subscribers worldwide. Messages can be received whilst a terminal is unmanned and subscribers are listed in world directories. The main service problem is the possibility of receiving terminals being in use or the international lines being engaged, which requires the operator to keep retrying to make the connection.

WIDE AREA NETWORKS 143

The usual way to provide access is by installing a Telex gateway on a local area network or conventional computer network. The gateway takes over the call connection part of the transfer and can perform functions such as repeated retries. Messages can be queued in a buffer, removing the problems of busy receiving terminals or engaged international circuits.

Although the Telex service meets a communication need, the preparation of documents by word processors has created the need for an enhanced document transmission service. There are however few agreed standards for word processor communications, characters or function codes. This is emphasised by the fact that only about three per cent of word processors are bought for the purpose of communicating. Because of the rapid advances in document preparation and the development of electronic office technology, a new service, Teletex, has been introduced.

Teletex is also the subject of international agreements and has a number of advantages over the Telex system. The most important are the extended character set – allowing graphical characters to be transferred – and the lack of restriction on the type of network used. British Telecom, for example, are providing a Teletex service on both the Public Switched Telephone Network (PSTN) and the Packet Switched Service (PSS). The service can therefore offer transmission speeds comparable with other local and remote data services, a considerable improvement over Telex.

The communications protocols for Teletex are relatively complicated and the technology used by the terminals is advanced. This has led to considerable delays in the introduction of the service, largely due to problems of approval. Because of this there are no Teletex gateways available at the present time, and relatively few standalone terminals have been approved. There are no technical reasons preventing the design of a Teletex gateway, and the need is certain to grow.

A full description of the Teletex facilities, applications and future growth is given by Price (1982).

Other public data services which can be accessed via the public telephone network or the packet switched service include mailbox services (eg Telecom Gold via PSS) and Prestel (via the PSTN).

Mailbox services are expected to rival the telephone in their importance to business users by the 1990s. The microcomputer user can access the Telecom Gold service, which is linked to the packet switched service, by setting up a modem link to a British Telecom PAD over the public telephone network. The use, design, installation and management of mailbox systems is described by Wilson (1983).

Prestel, the UK's viewdata service, is a well established general purpose information entry and retrieval system. The usual method of access is to use a modified colour television receiver connected to a normal telephone line. A more flexible connection is possible using a microcomputer with a Prestel adaptor, connected through the PSTN via a modem. As an example of a Prestel adaptor, Owl Micro-Communications market a plug-in card for the Apple II microcomputer which combines the Prestel functions and a 1200/75 baud modem for direct connection to a normal telephone line.

Another important class of public services are known as value added networks (VANs). Services such as Tymenet and Euronet rent basic communications lines from the PTT and provide PADs and other network devices such as protocol convertors. This allows a user to build up a communications network without the overheads of purchasing and maintaining the network. The suppliers of VANs also offer additional facilities such as time-shared computing resources and information retrieval systems to further enhance the value of their service.

The PTTs themselves have become aware of the importance of value added network services, and the Viewdata and Gold services are examples of British Telecom moving into this market. In a strategy survey by Maptek Europe (1984) it was estimated that more than two thirds of personal computer users will use value added networks. A particularly high priority was given to PC-to-PC communications when they are geographically dispersed.

8 Problems in Communications

SOME OF THE PITFALLS

Communications can mean anything from the link between two microcomputers sitting on the same desk to the global network of a large multinational company. As data may pass through a number of different hardware interfaces with routeing and other network services under software control it is not surprising that a wide variety of problems can be encountered.

Some of these problems will now be examined, and where possible, solutions suggested. These problems are not unique to microcomputer users connected to a computer network, they arise mainly due to incompatibilities between different vendors' equipment, to deficiencies in standards or in the network design. The problems were highlighted during interviews with a number of different computer network managers or communications specialists. It is therefore unlikely that all the problems would occur together in any one network.

SYNCHRONISATION DURING INTERACTIVE PROCESSING

This problem arises because of data buffering in the network and is most commonly experienced whilst using a host computer editor or on-line debugging aid. The problem shows itself as an apparently slow response to keyboard commands. The response to any keyboard command cannot affect the data already in transit, so this will appear on the screen before the command takes effect. The keyboard and screen display therefore appear to be out of step, which can be thought to be the result of a slow response time. As a simple example, if a file is being listed, and a control code is sent to the host to halt the listing, many lines of the file may be sent to the screen before the command to halt the listing takes effect.

The simple solution to this problem is to change the method of working and use the microcomputer as a local editor, with file transfer facilities set up between the microcomputer and the host computer. This is, after all, a natural development in distributed computing. Using a mainframe editor over a communications network is inefficient in both computer resources and communications overheads. When the problem is unavoidable however, it may be that the wrong transmission mode is being used for the application. In the discussion on PADs, a choice of transmission modes is usually provided, offering essentially line at a time or character at a time transfers. With the right choice selected, data buffering can be minimised.

FLOW CONTROL AND CHARACTER LOSS

Whenever two machines are linked together, one is likely to be able to transmit data faster than the other can receive it. This is not simply a question of setting the transmit and receive clocks to the same speed. Both transmission and reception of data take a finite time, as files are held on disk and must be accessed and moved to and from memory. Because of this a means of controlling the flow of data is essential unless the transmission speed is so slow that the receiving process never falls behind.

Several different methods of flow control are in use. The simplest technique, used in asynchronous communications, makes use of the V.24 control lines. Data Terminal Ready (DTR) is the circuit normally used by the receiving interface to indicate when it is free to receive data. This circuit is connected to the Data Set Ready (DSR) circuit of the transmitting interface.

An alternative flow control technique uses special control characters embedded in the data stream to control data flow. The main advantage of this technique is that no extra transmission circuits are needed other than the transmit and receive data circuits.

These techniques are quite adequate for minicomputer terminal networks, where they are used extensively, but are not always successful when applied to wide area networks. Again the problem arises because of data buffering, which can lead to the flow control characters being out of step with the data flow. If the receiving terminal sends the control code for stop transmission (X-OFF) when the internal buffer is nearly full, there can be data in transit which will cause buffer overflow and character loss before the X-OFF control code can take effect. This can only arise when end-to-end flow control is used, and most networks allow link-by-link flow control which overcomes this difficulty. Another source of difficulty with flow control is that data transparency is lost if the flow control characters can occur in the data stream. A data link protocol which converts data to printable characters would be required to transfer binary and non-text files.

This type of problem illustrates the difficulty in trying to interconnect systems using different protocols. The long term answer is to move towards a standard network architecture with a completely compatible set of protocols.

NON-TRANSPARENCY OF CIRCUIT

This problem has been illustrated in the previous sections, and arises because of the need to communicate with operating systems and control intelligent network devices. As the basic transmission unit is seven or eight bits long, data and control codes cannot be distinguished. There is also considerable variation in the options available on the serial interfaces found in equipment in use at the present time. Some interfaces allow only seven-bit data words plus parity, others allow five- to eight-bit data words plus parity. The parity bit can be odd, even, mark, space or ignored. Other complications can be found such as flow control available for seven-bit data words but not for eight-bit data words.

Whilst asynchronous devices with only the elementary X-ON/X-OFF flow control protocol can be connected to a computer network, the solution to data transparency problems is again to use a link level protocol designed for network applications. Protocols such as HDLC have flow control inherent in their design as data is transferred in packets, controlled by 'handshakes' between the two devices. Following on from this, by placing the data part of the transmission in a defined packet structure, the control function is separated from the data transfer function and there need be no restrictions placed on the data contents.

DEFICIENCIES IN SCREEN FORMAT OR RESOLUTION

An example of this type of problem has already been examined in the terminal emulator section. A microcomputer with a 24-line display format was tested as a replacement for an original terminal which has a 25-line display. In this case the problem was solved by providing the microcomputer with a 24-line window into the 25-line display. Other examples include the use of 40 characters per line displays in place of the 80 character displays in normal use. Again a window solution is used to provide a display of the whole screen, but this is less successful than the previous example.

The views of experienced users show that the screen format limitation is not a great problem for interactive applications using a scrolled display. However in applications using formatted screens, which make use of the whole screen, any screen limitations can be a serious handicap.

Another display limitation which can arise in terminal emulations is a reduction in character quality or graphics resolution compared with the original terminal. This is most likely to arise when a domestic quality microcomputer, with low bandwidth display and limited display memory, is used in place of a higher quality professional graphics terminal. The choice is simply one of price against performance.

CHARACTER SET PROBLEMS

Although an internationally agreed character set, ISO 646, is in almost universal use for microcomputers, there is still scope for some problems due to national variations. One common problem is the replacement of the pound sign (£) required for the British market by the dollar sign ($) used in the American market. Fortunately most foreign suppliers now customise the keyboard and character generator for the UK market.

The introduction of visual display unit terminals (VDUs) exposed the limitations of the seven-bit code, and methods of extending the code were developed. This is because of the greater flexibility in characters and functions that can be achieved with a VDU compared with the earlier electromechanical Teletype device. Two international standards, ISO 2022 and ISO 6429, cover code extension techniques for 7-bit and 8-bit character sets, and additional control functions for VDU devices. Definition of character sets is a complex problem, and these standards have the form of limited conformance, which means appropriate subsets can be selected. The present position is that individual users and suppliers have developed their own graphical character subsets, which has created a problem of compatibility. It seems likely that a number of graphics sets will emerge to serve different purposes.

Another problem associated with character sets is the continuing use of 40 characters per line microcomputers for business applications. These machines were intended for the domestic market and feature a limited upper-case only display. Upper- and lower-case can be stored internally however, and output to printers and to other devices over communications links can include the full upper- and lower-case character set. Various links are used to convey the case of characters on the screen, such as displaying all upper-case characters in inverted video.

KEYBOARD LIMITATIONS

Although the keyboard is still the most important input device, it is surprising how wide a variation can be found in keyboard layout and characters supported. The basic QWERTY layout of the typewriter keyboard is universally accepted, but there is no uniformity in positioning such basic control keys as carriage return, line feed and the shift keys. The positioning of the punctuation and special characters also seems to follow no logical rules and characters are often missing from the keyboard. Some missing characters can be entered from the keyboard, but only by a combination of control and shift keys that bear no resemblance to the desired character.

There is an opposing view to the need for a 'standard' that regards the needs of users to be so varied that there is no purpose in defining a standard keyboard.

The standards approach has been to use the concept of harmonisation. With this, a standard specifies precisely the commonly used keys, and leaves flexibility to accommodate the graphics and control functions needed on a national basis. However, little progress has been made so far in applying harmonisation or even preferred allocations to the keys for the control functions. Full details are given in the NCC Guide to Computing Standards, Number 17.

As keyboard skill is an acquired skill, some weight should be put on keyboard layout when choosing replacement equipment, especially if a number of different suppliers are being considered.

INCOMPATIBLE CONTROL CODES

The first problem associated with character sets was in the display of characters, either occurring as differences in the displayed character for a specific code or as a restriction in the characters available (eg upper-case only). The second problem is concerned with the function of the control characters and control sequences. As the control characters have a direct affect on applications or operating systems, these problems are potentially more serious.

Terminals and computers are designed to a common specification, and terminal interfaces, controllers and the keyboard monitor part of the operating system, have individual design peculiarities. Leaving aside synchronous terminals for the moment, asynchronous terminals can usually be swapped but a wide variety of operational difficulties can occur. Some of the more frequently encountered differences are in the delete function, tab function and action on keying carriage return. The more successful terminals, for example the VT100 from Digital Equipment, can be configured in a variety of ways from the keyboard. Because these advanced terminals offer features which considerably enhance interactive data entry, applications often make use of the terminals' extended control features. One of the most commonly used features is direct cursor addressing for the creation of forms-based data entry systems. If an emulation is to work satisfactorily in place of an original terminal it must have a full implementation of the extended control functions. Often the only way to find out if a replacement terminal can be used successfully is by carrying out a field trial, as a comparison of specifications can be difficult.

Host computers using synchronous protocols and terminals present a quite different problem. As each supplier's protocols and terminal features are quite different, terminals cannot be swapped in the same way that asynchronous terminals can. Emulations can be written for microcomputers however, with varying degrees of success. The main point to watch out for is the restriction in facilities offered by an emulation, which may or may not be important for particular applications. These restrictions are usually forced onto the writer of the emulator software because the power of the microcomputer is inadequate to support all functions. The best advice that can be given is to ask for a thorough field trial.

POOR RESPONSE TIMES AND EXCESSIVE DELAY IN KEYBOARD ECHO

When using a microcomputer or terminal interactively, the user needs an immediate display of characters typed as visual feedback. Some proficient users can type ahead of the display, especially when carrying out repetitive control tasks, but when the display needs to be monitored the timing of keyboard echo is critical. Delays greater than 200 ms are unacceptable, and the problem is accentuated as the delay is increased. A particularly disturbing effect occurs when the echo delay is random and characters are echoed in unpredictable bursts.

The second type of problem is excessive delay in the response to a command. Users find a delay of four to five seconds acceptable, but as delays increase beyond this, user frustration rapidly builds up. The microcomputer user is used to very good response times, because as the sole user of the machine there are no other users competing for the machine's resources. Microcomputer applications are usually much more responsive than host computer systems, particularly in changing screen displays.

A degradation in response times is the inevitable result of the loading of shared systems. It is probably the most common cause of complaints about computer systems, and can usually only be solved by spending money on extra resources. There are however some steps that can be taken to improve the situation in certain circumstances.

One example has already been quoted in reference to the use of host computer editing facilities. This is an example of failing to take advantage of distributed computing. Text editing is a considerable drain on both host computers and communications network resources. The better solution is to move files between the host and the microcomputer and to use local editing facilities.

The way keyboard echo is obtained can also be examined. Keyboard echo to a terminal group connected to a minicomputer host is usually provided by the host computer. The advantage of this technique is that it gives feedback that the transmission is error free and that the host is on-line. In an extensive computer network however, delays are inevitable and local echo should be used. The second point to note is that in a distributed computing network applications should run on a transaction basis rather than character-by-character. This means that data transfers will be line-by-line or screen-by-screen, and in many applications complete files will be transferred. There will be no direct interaction between the user and the host computer, and all keyboard echo and line editing functions will be carried out locally.

It is very difficult to be specific about the causes of poor response times as these are many and varied. One important factor did emerge from interviews with users: the source of problems could usually be isolated to the computing resources connected to the network rather than the network itself. In local area networks, for example, the file server is often the source of bottlenecks. Difficulties with gateways have also been reported. Most local area networks are being run well below their capacity and so give a good performance. Wide area networks however are more prone to problems as high error rates can reduce the throughput dramatically and continuous monitoring is required to maintain a good service.

9 Conclusions

Microcomputers have revolutionised the computer industry by making computing accessible to the manager or office worker. This success is largely due to a small number of software packages including word processing, spreadsheets and data base systems. All of these software families were originally developed as stand-alone systems. Very often however the information or document needed is held on another machine and unless the machines are linked they are very likely to end up as expensive executive toys.

In this book the attractive possibilities for the microcomputer as a communications device have been described. Even the most basic microcomputers sold at the present time can be provided with useful communications facilities. This is mainly because of the almost universal provision of a serial interface and the wide availability of software designed to use this port. With this simple interface the microcomputer can become a terminal to a variety of host computers, can transfer files to and from other machines, and can share the resources of a computer network. By connecting to the telephone network, electronic messages can be sent and received on a global scale.

The steps to be taken in making the physical connection, and selecting and operating communications software have been described in detail to help the end user overcome the technical difficulties that can be encountered.

The earlier 8-bit microcomputers were limited in processing power and memory capacity making the handling of communications protocols difficult, which restricted the communications support that could be provided. This has now changed with the development of more powerful 16-bit machines, and the trend towards greater performance is still continuing.

Networking microcomputers has become a major development task for many organisations as the importance of communications is realised. Networks can extend from a local area network covering a single office or site up to the global scale of a wide area network. The methods of interfacing to networks have been described by referring to some representative products. These have also been used to illustrate the kind of services available to the network user. The key to successful networking and interworking lies with standards. The ISO model provides a framework for a set of protocols covering the needs of a communications network from electrical standard to the display of data on a screen.

Progress is being made from the bottom up, so that physical, electrical and data link protocols are well defined, but the higher levels, for example the specification of control codes embedded in a word processor text file, remain to be defined. Standards at the higher levels, the presentation and applications layers, will take longer to define and implement because of the variety of application and services they need, and the complexity of the problem. Also, work was started on the lower layers first as these must be in place before the higher layers can operate. When the complete set of standards has been defined it will be possible to transfer files and documents between machines with a good chance of being able to use them with existing software packages.

APPENDIX 1

References and Bibliography

Bleazard, G. B., *Handbook of Data Communications*, NCC Publications, 1982

Bleazard, G. B., *Why Packet Switching?*, NCC Publications, 1979

Da Cruz, F. and Catchings, W., Kermit: A File-Transfer Protocol for Universities,
   Part 1, *Byte*, June 1984, pp 255-278,
   Part 2, *Byte*, July 1984, pp 143-403

Digital, Intel and Xerox,
   The Ethernet: A Local Area Network,
   Data Link Layer and Physical Layer Specifications,
   Digital Equipment Corporation, Intel Corporation and Xerox Corporation, 1980

Flint, D. C., *The Data Ring Main*, John Wiley & Sons, 1983

Gee, K. C. E., *Local Area Networks*, NCC Publications, 1982

Gee, K. C. E., *Local Area Network Gateways*, NCC Publications, 1983

Gee, K. C. E., *Proprietary Network Architectures*, NCC Publications, 1981

Healey, M., *Managers' Micros*, NCC Seminar, Sheffield Polytechnic, 4-5 July 1984

Hogan, T., *Osbourne CP/M User Guide*, Osbourne/McGraw-Hill, 1982

Maptek Europe, *Special Strategy Conference on USA Markets and Technology,* London, May 1984

McGregor Ross, H., Keyboard Layouts, *Guides to Computer Standards,* Number 17, The National Computing Centre, November 1980

McNamara, J. E., *Technical Aspects of Data Communications,* Digital Press, 1977

Price, S. G., *Preparing for Teletex,* NCC Publications, 1982

Scott, P. R. D., *Modems in Data Communications,* NCC Publications, 1980

Shoch, J. F. and Hupp, J. A., *Performance of an Ethernet Local Network,* A Preliminary Report, Local Area Communications Network Symposium, Boston, May 1979

Wenig, R. P., Pardoe, T. D. and Erb, D. P., *Data Communications for Microcomputers,* International Management Services Inc Seminar presented by Frost & Sullivan Ltd, 1983

Wilson, P. A., *Introducing the Electronic Mailbox,* NCC Publications, 1983

APPENDIX 2

Suppliers of Communications Software and Equipment

Act (Pulsar) Ltd, Freepost, Birmingham, B16 1BR. 021-455-7000.
  Asynchronous communications software for Apricot, Sirius, IBM PC and Apple II.

Adabas Software Ltd, Laurie House, 22 Colyear Street, Derby, DE1 1LA. 0332-372535.
  IBM PC to IBM host link.

Advanced Quality Software, 88 St Benedicts Street, Norwich, NR2 4AB. 0603-21117.
  Apple communications software packages.

Alpha Data Systems Ltd, 114 Ashley Road, St Albans, Herts, AL1 5JR.
  Communications equipment.

Amazon Computers Ltd, Linford Wood Business Centre, Sunrise Parkway, Milton Keynes, MK14 6LQ. 0908-664123.
  Micro-mini-mainframe communications software for wide range of machines.

Anderson Jacobson Ltd, 752 Deal Avenue, Slough, Berks, SL1 4SJ. 0753-25172.
  Communications equipment.

Beauforts Computer Systems, 22 Highfield Road, Edgbaston, Birmingham, B15 3DP. 021-454-7467.
  IBM PC to IBM host link.

Boeing Computer Services, 19 Fitzroy Street, London, W1P 5AB. 01-631-0808.
   Asynchronous terminal emulator for IBM PC.

Browns Operating System Services Ltd, Westminster Bank Chambers, Blackheath, London, SE3 9RD. 01-852-3299.
   IBM 3270 emulations for range of micros, protocol converters.

Camtec Electronics Ltd, Melton Street, Leicester, LE1 3NA. 0533-537534.
   Communications equipment.

Codex (UK) Ltd, 114/116 Thornton Road, Thornton Heath, Surrey, CR4 6XB. 01-689-2101.
   Communications equipment.

Compushack, 12 Nottingham Place, London, W1M 3FA. 01-935-0480.
   Asynchronous terminal emulation for IBM PC and compatibles.

Computer and Systems Engineering PLC, Caxton Way, Watford Business Park, Watford, Herts, WD1 8XH. 0923-33500.
   Communications equipment.

Cortex Computer Systems, Cortex House, 5 Union Street, Bedford, MK40 2YR. 0234-217721.
   Asynchronous communications software for Commodore PET. IBM 2780/3780 Cluster Controller for PET.

CPS (Data Systems) Ltd, Arden House, 1102 Warwick Road, Acocks Green, Birmingham, B27 6BH. 021-707-3866.
   IBM 3270 emulation for IBM PC. Agent for ABM Computer Systems.

Cullinet Software Ltd, The Broadway, Stanmore, Middlesex, HA7 4D9. 01-954-7333.
   IBM PC to IBM host link.

Dacoll Ltd, Dacoll House, Kingsbury Road, Minworth, Birmingham, B76 9DF. 021-351-5451.
   Terminal emulators for IBM, ICL, Univac and Honeywell.

SUPPLIERS

Data Logic Ltd, 29 Marylebone Road, London, NW1. 01-486-7288.
Communications equipment.

Dataview Systems, Portreeves House, East Bay, Colchester, Essex, CO1 2XB. 0206-865835.
Honeywell VIP 7800 terminal emulator for DEC Rainbow and IBM PC.

Davidson Richards, 28/29 Charwood Street, Derby, DE1 2GU. 0332-383231.
Supplier of wide range of micro communications software.

Derwent Data Systems Ltd, 18 Norfolk Street, Sunderland, Tyne & Wear, SR1 1EA. 0783-652026.
Asynchronous communications software for CP/M, CP/M 86 and MS-DOS machines.

Encotel, 7 Imperial Way, Croydon, Surrey. 01-686-9687.
Synchronous (IBM) and asynchronous terminal emulations for CP/M and CP/M 86 machines.

Formscan Ltd, Apex House, West End, Frome, Somerset. 0373-61446.
Protocol converters.

Gandalf Digital Communications Ltd, 19 Kingsland Grange, Woolston, Warrington, WA1 4RW. 0925-818484.
Communications equipment.

General Computer Systems (UK) Ltd, Orion Park, 226-236 Northfield Avenue, London, W13 9QU. 01-579-9401.
Communications controller for cluster of Apples to IBM host.

General Data Communications (UK) Ltd, Toutley Road, Wokingham, Berkshire, RG11 5QN. 0734-791444.
Communications equipment.

IAL Data Communications Ltd, Jays Close, Wiables, Basingstoke, Hampshire, RG22 4BY. 0256-59222.
Communications equipment.

Informatics General (UK) Ltd, Africa House, 64-78 Kingsway, London, WC2B 6AL. 01-242-0770.
IBM PC to IBM host links.

Infotron Systems Ltd, Systems House, Poundbury Road, Dorchester, Dorset, DT1 1TA. 0305-66016.
Communications equipment.

Intelligence Ireland, Nagor House, Dundrum Road, Windy Arbour, Dublin 14, Ireland. Dublin 988555.
Asynchronous and synchronous (IBM) terminal emulations for CP/M and CP/M 86 machines.

Jaguar Communications Ltd, Elton House, London Road, St Albans, Herts. 0727-65129.
Range of terminal emulations.

Jardgate Ltd, 197-213 Lyham Road, Brixton, London, SW2 5PY. 01-671-6321.
Variety of synchronous (IBM, ICL) terminal emulations for CP/M 80 and CP/M 86 machines, communications equipment.

Logica VTS Ltd, 84 Newman Street, London, W1A 4SE. 01-637-7761.
IBM PC to IBM host, and PC to PC communications software.

Master Systems (Data Products) Ltd, 100 Park Street, Camberley, Surrey, GU15 2NY. 0276-28527.
Communications equipment.

Mator Systems Ltd, 134-140 Church Road, Hove, Sussex, BN3 2DL. 0273-726464.
Wide range of synchronous and asynchronous communications software.

MDS Computer Systems (UK) Ltd, 84 Upper Richmond Road, London SW15 2ST. 01-874-6404.
IBM 3270 terminal emulator.

Metascybe Systems Ltd, 78 Long Lane, London, EC1A 9ET. 01-606-6865/6.
Supplier of protocol converters, communications software.

Micom Borer Ltd, 15 Craddock Road, Reading, Berkshire, RG2 0JT. 0734-866801.
Communications equipment.

SUPPLIERS

Micro Trend UK, PO Box 51, Pately Bridge, Harrogate, North Yorkshire, HG3 5DF. 0423-711878.
IBM 3780 terminal emulation for CP/M machines.

MOM Systems, Bon Accord Street, Aberdeen, Scotland. 0224-571825.
Asynchronous/bisynchronous communications software for CP/M, CP/M 86, CROMIX or MS DOS machines.

MPI International, 8 Cambridge House, Cambridge Road, Barking, Essex. 01-591-6511.
Supplier of communications software.

MSA, MSA House, Cedars Road, Maidenhead, Berkshire, SL6 1SA. 0628-39242.
IBM PC to IBM host links.

Network Designers Ltd, The Old Berkshire Hunt Kennels, Lingston Bagpuize, Oxon, OX13 5AP. 0865-821177.
Synchronous (ICL) terminal emulation for wide variety of machines. Asynchronous communications software for 16-bit machines.

Nolton Communications Ltd, Fieldings Road, Cheshunt, Herts, EN8 9TL. 0992-33555.
Communications equipment.

Nova Micro Centres, 7 Imperial Way, Croydon Airport Industrial Estate, Croydon, CR0 4RR. 01-681-8620/8633.
Synchronous (IBM) terminal emulations for IBM PC and compatibles.

Owl Micro Communications, The Maltings, Station Road, Sawbridgeworth, Herts. 0279-723848.
Communications equipment and terminal emulators for Apple micros.

P & P Micro Distributors, New Hall Hey Road, Rossendale, Lancs, BB4 6JG. 0706-217744.
Wide range of communications software.

Racal-Milgo Ltd, Landata House, Station Road, Hook, Hants, RG27 9NF. 025672-3911.
Communications equipment.

Root Computers, Saunderson House, Hayne Street, London, EC1A 9HH. 01-726-6501.
Supplier of communications software for UNIX machines.

Sapphire Systems, 1-3 Park Avenue, Ilford, Essex. 01-554-0582.
Asynchronous communications software for CP/M machines.

Scicon Computer Services, Brick Close, Kiln Farm, Milton Keynes, Bucks, MK11 3EJ.
Communications equipment.

Techland Systems International Ltd, Wyebridge House, Cores End Road, Bourne End, Bucks, SL8 5HH. 06285-26535.
Synchronous (IBM 3270) and asynchronous (UT100) terminal emulations for IBM PC.

Tech-Nel Data Products, 8 Haslemere Way, Banbury, Oxon, OX16 8TY. 0295-65781.
Communications equipment.

Telesystems Ltd, The Geans, 3 Wycombe Road, Prestwood, Bucks, HP16 0NZ. 02406-6365.
Supplier of asynchronous communications software for wide range of machines.

Thorn EMI Datatech, North Feltham Trading Estate, Feltham, Middx. 01-890-1477.
Communications equipment.

Timeplex, Timeplex House, North Parkway, Leeds, LS14 6PX. 0532-735141.
Communications equipment.

Tradesoft, Tradesoft House, 62 Weir Road, Wimbledon, London, SW19 8UG. 01-879-1144.
Suppliers of wide range of communications software.

Transaction Technology, 77 Peascod Street, Windsor, Berks, SL4 1OJ. 07535-56789.
Asynchronous micro communications software for wide range of 8-bit and 16-bit micros running under CP/M, CP/M 86 and MS-DOS. Asynchronous micro-mainframe link.

Transam, Micro Systems Limited, 59/61 Theobalds Road, London, WC1X 8SF. 01-404-4554.
Asynchronous terminal emulator for HX-20 micro.

SUPPLIERS

Ultracomp Ltd, 2 Candle Ford Close, Bracknell, Berks, RG12 2JZ. 0344-426024.
  Apple to ICL C01/C03 packages.

VHA Computer Services, Coal Road, Leeds, LS14 2AL. 0532-737475.
  ICL C03 emulators for Act Sirius and Apricot micros.

Virtual Micro Systems, 20 Queens Road, Reading, RG1 4AL. 0734-599408.
  Micro/mini communications for wide range of machines.

West Surrey Computers Ltd, Chandler House, Anchor Hill, Knaphill, Woking, Surrey, GU21 2NL. 04867-88561.
  IBM PC to IBM host links.

Xoren Computing Ltd, 28 Maddox Street, London, W1R 9PF. 01-629-5932.
  File transfer and terminal emulation for PDP11 computers.

Xtec Limited, High Street, Hartley Wintney, Basingstoke, Hampshire, RG27 8PB. 025126-4222/4233/4344.
  ICL and IBM protocol converters.

APPENDIX 3

Abbreviations

ACK	Acknowledge
ALU	Arithmetic and Logic Unit
ANSI	American National Standards Institute
ASCII	American Standard Code for Information Exchange
BIOS	Basic Input Output System
BIT	Binary Digit
bps	bits per second
BS	Backspace
CAP	Cable Access Point
CATV	Community Antenna Television
CCITT	The International Telegraph and Telephone Consultative Committee
CCT	Circuit
CD	Carrier Detect
CDSL	Connect Data Set to Line
CP/M	Control Program for Microcomputers
CPU	Central Processing Unit

CR	Carriage Return
CRC	Cyclic Redundancy Check
CSMA/CA	Carrier Sense Multiple Access with Collision Avoidance
CSMA/CD	Carrier Sense Multiple Access with Collision Detection
CSV	Comma Separated Value
CTS	Clear to Send
DCE	Data Communication Equipment
DIF	Data Interchange Format
DIL	Dual-In-Line
DP	Data Processing
DSR	Data Set Ready
DTE	Data Terminal Equipment
DTR	Data Terminal Ready
EBCDIC	Extended Binary Coded Decimal Interchange Code
EIA	Electronic Industries Association
EMAS	Edinburgh Multi-Access System
EOF	End of File
EOT	End of Text
HDLC	High Level Data Link Control
HEX	Hexadecimal
Hz	Hertz
IC	Integrated Circuit
IEEE	Institute of Electrical and Electronics Engineers
I/O	Input/Output
ISDN	Integrated Services Digital Network

ABBREVIATIONS

ISO	International Standards Organisation
JANET	Joint Academic Network
KHz	Kilo Hertz
LAN	Local Area Network
LED	Light Emitting Diode
LF	Line Feed
LST	List
MHz	Mega Hertz
MNP	Microcom Network Protocol
NAK	Negative Acknowledge
OSI	Open Systems Interconnection
PABX	Private Automatic Branch Exchange
PAD	Packet Assembler/Disassembler
PC	Personal Computer
PG	Protective Ground
PIP	Peripheral Interchange Program
PTP	Paper Tape Punch
PTR	Paper Tape Reader
PSS	Packet Switch Stream (UK Public Packet Switched Service)
PSTN	Public Switched Telephone Network
PTT	Postal, Telegraph and Telephone Authority
PUN	Punch
RAM	Random Access Memory
RD	Received Data
RDR	Reader
RI	Ring Indicator

ROM	Read Only Memory
RT	Receiver Timing
RTS	Request to Send
SCD	Secondary Received Line Signal Detector
SCT	Secondary Clear to Send
SDLC	Synchronous Data Link Control
SG	Signal Ground
SNA	Systems Network Architecture
SOH	Start of Heading
SP	Space
SRD	Secondary Received Data
SRT	Secondary Request to Send
ST	Send Timing
STD	Secondary Transmitted Data
TAP	Terminal Access Point
TD	Transmitted Data
TTY	Teletype
VAN	Value Added Network
VDT	Visual Display Terminal
VDU	Visual Display Unit
WAN	Wide Area Network

Appendix 4

Glossary of Communications Terms

Access method	The mechanism used to share a common physical transmission medium between a number of users.
Analogue transmission	Transmission of a continuously variable signal.
Applications layer, OSI	The highest layer defined by the ISO Reference Model concerned with providing services to applications programs which exchange information with others.
Arithmetic checksum	Character or characters added to a block of data based on some arithmetic property of this block which is used for error detection.
Asynchronous transmission	Transmission in which there is no fixed time interval between characters. The start of a character is marked by a start signal and the character is terminated by a stop signal.
Bandwidth	The range of frequencies available for signalling in a communications channel.
Baseband signalling	Transmission of a signal at its original frequencies, ie unmodulated.

Baud rate	The unit of discrete signalling speed per second; the modulation rate. The baud rate is equivalent to bits per second only if each signal represents exactly one bit.
Bit	Abbreviation of binary digit. The signal element of transmission in binary notation which can take the two values '0'(OFF) or '1'(ON).
Bit rate	The number of bits transferred in unit time, usually expressed in bits per second (bps).
Break-out box	A device which can be inserted in a serial transmission line to enable the signal levels to be monitored and cross-connections to be made.
Bridge	A device used to connect similar networks together.
Broadband signalling	Transmission of a signal at a frequency higher than the original, ie modulated. Many signals will usually be carried by the same transmission medium by occupying separate frequency bands.
Buffer insertion	(*See* Register insertion)
Bus	A single common data highway shared by a number of devices.
Byte	A sequence of bits, normally eight, which represents one character.
Byte stuffing	A method of obtaining data transmission transparency by inserting special prefix characters to mark modified characters.
Carrier sense multiple access (CSMA)	A method of sharing a common data highway by testing whether the highway is in use before transmission. To

GLOSSARY 171

	avoid the problem of two stations trying to transmit at the same time, collision avoidance (CSMA/CA) or collision detection (CSMA/CD) techniques are usually adopted.
Central processing unit (CPU)	The central part of a computer which includes all the major control and arithmetic units, with the exception of memory and peripheral interfaces.
Character set	The set of letters, figures, punctuation and control codes in a message. Each character is usually represented by one byte.
Check bit or check character	A bit or character associated with a character or block of characters for error detection purposes.
Circuit switching	Conventional interconnection where a two-way fixed bandwidth circuit is allocated exclusively to the parties concerned for the duration of the call.
Concentrator	A device that provides communications facilities between a number of low-speed devices and one or more high-speed channels.
Concurrency	Availability of a number of services or applications at the same time.
Connectivity	The flexibility of a network in providing connections to a number of different systems.
Contention	An access method which relies on competition for a shared data highway.
Control codes	A non-printable character whose function is to initiate, modify or stop a control operation.

Cyclic redundancy check	A method used to detect errors in transmitted data by adding a string of specially constructed check characters to the data.
Data base	An organised collection of data available to a number of users.
Data buffering	The storage of data characters for speed matching or interfacing purposes.
Data communications equipment (DCE)	In data networks, a DCE is any device used to interface the data terminal equipment (DTE) to the communications line.
Data link layer, OSI	The layer defined by the ISO Reference Model concerned with ensuring transmission errors across a data link are detected and corrected.
Data link protocol	A set of rules used to ensure that transmission errors across a data link are detected and corrected. High Level Data Link Control is a typical example.
Data switch	A circuit switching system designed to carry data exclusively.
Data terminal equipment (DTE)	In a data network, a DTE is any device attached to the network which originates data or is a destination device for data.
Digital transmission	This is the transmission of data characters by coding into discrete signals.
Distributed computing	An arrangement of computing facilities where a number of computers are spread around a site with high-speed interconnecting links for resource sharing.

GLOSSARY

Echo — The display of a character keyed in at a terminal. The displayed character may be routed from the keyboard to the screen locally or may be transmitted from the remote computer.

Editor — A program, usually supplied with the computer, for the creation of text files.

Electronic mail — A service for the transfer of documents between computer users attached to a network.

Empty slot — (*See* Slotted ring)

Encryption — The encoding of data into a new form which needs a specific algorithm to decipher it.

Error rate — The probability, within a given sample size of bits, characters or blocks, of one being in error.

Field — Part of a group of characters with special significance.

File — A logical organisation of characters or data, stored on the computer memory, disk, tape or other storage medium.

Filter — Another name for a bridge, a device for linking similar networks.

Fixed slot — A network access method which gives each node on a ring a slot for its exclusive use.

Floppy disk — A computer storage medium which uses a thin flexible disk, usually 8 inches or $5\frac{1}{4}$ inches in diameter, to store digital data.

Flow control — A mechanism for ensuring that characters can be exchanged between two devices in an orderly manner without loss.

Frame	A group of characters containing several fields which form a single message.
Frequency division multiplexing	A method of sharing a transmission line by dividing the available bandwidth into a number of smaller bandwidth channels.
Front-end processor	A special-purpose computer dedicated to handling the communications needs of a number of terminals connected to a host computer.
Full duplex	Simultaneous transmission of data between two devices in both directions.
Gateway	A device for interconnecting dissimilar networks.
Half duplex	Transmission of data between two devices in both directions, but not simultaneously.
Host	A central computer system which can run applications for users connected via a computer network.
Hunt group	Facilities which are accessed by a queuing mechanism.
Input/output device	Any device such as a keyboard or printer which can be used to enter data into a computer or read data from a computer.
Interchange circuit	The term used by the CCITT for interconnecting circuits at the interface between terminals (DTE) and modems (DCE).
Interrupt	A computer hardware feature which allows a low-priority operation to be

GLOSSARY

	temporarily suspended to enable a high-priority operation to be carried out.
Interworking	The ability to exchange data between computers from different suppliers.
Invertible file transfer	A file transfer from one machine to another and back again without loss of any attributes.
Leased line	A rented telephone circuit which provides a permanent connection between two points.
Local area network	A communications network limited to a small area, usually a single site.
Macro	A group of instructions or table of data which can be inserted in a program using a single name.
Mailbox	Part of the storage in an electronic mail service which hold the documents for one user.
Mainframe	A large computer system, designed to carry out high-speed data processing.
Manchester encoding	A signalling scheme which also carries clocking information.
Microcomputer	A small computer designed for personal use.
Micronet	A local area network for microcomputers.
Microprocessor	A large-scale integrated circuit providing the control and arithmetic functions which form part of a complete microcomputer.
Minicomputer	A medium-sized computer capable of supporting departmental needs with a number of interactive terminals.

Modem	Abbreviation for modulator/demodulator, a device which converts digital information into an analogue form suitable for transmission over the analogue telephone network. The modem also performs the reverse function.
Multi-drop	A network configuration in which the central system communicates successively with a number of attached terminals. (*See* Polled network)
Multiplexing	A mechanism for carrying a number of low-speed data signals over one high-speed channel.
Network layer, OSI	The layer defined by the ISO Reference Model concerned with the routeing and switching operations associated with the establishment and maintenance of a connection between systems.
Null-modem cable	A cable which allows two terminals to communicate over a V.24 serial line by simulating the presence of modems and a telephone line.
Open systems interconnections	Standardisation procedures which allow terminals, computers and other devices from different suppliers to exchange data.
Operating system	Software which provides all the basic services required to run applications programs, handle input/output devices and manage the storage medium for a computer.
Packet switching	A method of carrying data across a data transmission network in blocks

GLOSSARY 177

with a defined format containing control and data fields. Each packet carries information which allows the packet switching exchanges to route it to its destination without a permanent circuit being maintained.

PAD Abbreviation for packet assembler/disassembler. A device which forms the serial data stream from a simple terminal into packets for transmission across a packet switching network. The PAD also performs the reverse function.

Parity bit A non-information bit which is added to the bits forming a character to help detect transmission errors. The parity bit is added such that the resulting number of 1s in the group is odd (odd parity) or even (even parity).

Peripheral interchange program (PIP) A utility program which allows the transfer of files between the peripherals attached to a computer. A number of facilities are usually provided for the user such as converting upper-case letters to lower-case, or vice versa, and filtering form feeds from the source file.

Physical layer, OSI The layer defined by the ISO Reference Model concerned with the interface to the cable and the control of its use.

Point-to-point A network in which pairs of end points are connected by a direct link.

Polled network A type of network in which each channel is periodically interrogated to determine if it is active.

Presentation layer, OSI	The layer defined by the ISO Reference Model concerned with the two-way function of taking information from applications and converting it into a form suitable for a common, machine-independent understanding.
Proprietary network architecture	A network structure which has been designed by a single supplier, and is therefore under the supplier's complete control.
Protocol	A set of rules governing the format and exchange of data between two devices such that reliable communication is maintained.
Protocol converter	A device which enables communication between equipment using different protocols, by effecting conversion between the protocols.
Random access memory (RAM)	A type of memory element and structure which allows individual elements to be written to or read from.
Read only memory (ROM)	A permanent memory structure which can only be read from. Some types of read only memory (EPROM) can be reprogrammed using special equipment.
Record	The first major unit in a data base which can be uniquely identified. Each record contains a number of individual data values or fields.
Register insertion	A type of ring network access method which requires repeaters that have a buffer which can be switched in and out of the circuit as required.

GLOSSARY

Repeater	A device inserted into a transmission line to restore the waveform and amplitude of a signal.
Response time	The time between sending the last character of a message from a terminal to the receipt of the first character of the reply. It includes terminal delay, network delay, and host computer delay.
Serial interface	The circuitry which allows each bit in a string of characters to be sent sequentially along a single channel.
Serial transmission	The transmission of data in which each bit is transferred along a single channel sequentially rather than simultaneously as in parallel transmission.
Server	A device connected to a network which provides a service, such as printer spooling, to the devices attached to the network.
Session layer, OSI	The layer defined by the ISO Reference Model concerned with the transfer of data between two applications. The two applications form a liaison for this purpose, known as a session.
Signalling	The modification of some transmission characteristic to carry information.
Slotted ring	A network access method where one or more packets circulate continuously around a ring.
Speed matching	The connection of two devices operating at different baud rates.

Star network	A type of network topology which has a central hub with links radiating out from it.
Statistical multiplexer	An intelligent time-division multiplexer that has unequal time slots to maximise the utilisation of transmission capacity.
Synchronous transmission	A transmission technique in which data is sent without start and stop bits between characters at the start of transmission. A special group of characters is sent to synchronise the receiver with the incoming data stream.
Telephone circuit	A communications link, characterised by limited bandwidth (300 Hz to 3,400 Hz) and poor noise and interference properties, designed to carry speech traffic.
Teletex	A public document transfer service with a large character set.
Teletype	An early electromechanical computer terminal, operating at low speeds, with upper-case only characters.
Telex	A well-established public document transfer service with a limited character set and low transmission speed.
Terminal concentrator	A device which allows a number of low-speed terminals to share one or more high-speed channels.
Terminal emulator	Software which allows a microcomputer to perform like a proprietary computer terminal.
Time division multiplexing	A technique of dividing the transmission capacity of a line into time slots

	which can be allocated to a number of channels.
Token passing	A ring network access method which utilises a special packet or token that is passed from one user to another and grants the holder of it the exclusive use of the network.
Transmission mode	The kind of signalling used to carry data over a transmission line.
Transparency	A communications link is said to be transparent if it allows the transfer of all characters without altering the message or taking control action.
Transport layer, OSI	The layer defined by the ISO Reference Model concerned with the provision of a particular class and quality of transmission. The transport layer is responsible for optimising the available resources to provide this service.
Tree network	A type of network topology in which the data highways can branch a number of times before reaching the nodes.
V-series recommendations	The CCITT recommendations for data transmission over telephone (ie analogue) circuits.
Value added network	A network which provides services in addition to simple connections, such as protocol conversion and network management.
Videotex	Public information service which can be accessed via a modified television receiver or special terminal connected to an ordinary telephone line.

Virtual circuit — A connection between devices which uses a virtual circuit appears to be a permanent physical circuit to the users, but in reality it is shared by other users and may involve more than one physical route.

Wide area network — A communications network which can extend to a global scale, involving the use of circuits provided by various PTTs.

Winchester disk — A storage medium which achieves high storage density and data transfer rate by sealing rigid disks in an evacuated enclosure.

Word length — The size in bits of the internal computer registers and data highways.

Workstation — Equipment based on the microcomputer which provides a wide range of services for the electronic office environment.

X-series recommendations — The CCITT recommendations for data transmission over public data networks.

Index

access methods	98, **114**
ANSI	60
Apple	20, 63, 86, 144
applications layer, OSI	57, 90
arithmetic checksum	92
ASCII	61, 67, 68, 90
asynchronous transmission	**28,** 62, 71, 78, 146
auto dial/auto answer	73
baseband network	49, 113
baud rate	15, 35, 48
BBC Micro	62, 86
binary files	67, 70, 84
bit rate	28, 35
break-out box	47
bridge	99, 125
Bristol University Computing Centre	**85**
broadband network	24, 101, 113
broadcasting	104
BT Merlin	87
buffer insertion	117
bus network	98, 112, 115
Byrom Software	78
byte stuffing	**90**
cable	49, 99, 110
– coaxial	99, 100, 110, 112, 114, 118

- optical fibre   99, 100, 110, 112, 114
- ribbon   110
- twisted pair   99, 110
cable tape   118
Cambridge Ring   98, 117
Cannon D-type connector   36
carrier sense multiple access (CSMA)   116
carrier sense multiple access with collision avoidance (CSMA/CA)   116
carrier sense multiple access with collision detection (CSMA/CD)   116
CATV   113
CCITT   32, 51
central processing unit   15, 17, 27
character   30, 67, **88**
character set   61, 68, 148
Cifer   86
circuit switching   130, 140
coloured books (rainbow books)   134
comma separated value (CSV)   95
Commodore   86
concurrency   20, 21, 73
conferencing   125
connectivity   98
contention   102, 115
control character   38, 61, 70, 77, 85, 150
CP/M   75, 86
CPU
    (*see* central processing unit)
CSMA
    (*see* carrier sense multiple access)
CSMA/CA
    (*see* carrier sense multiple access with collision avoidance)
CSMA/CD
    (*see* carrier sense multiple access with collision detection)

cyclic redundancy check (CRC)	92
Dacoll	65
data base	60, 106
data buffering	135, 145, 147
data communications equipment (DCE)	30, **33**
data interchange format (DIF)	95
data link layer, OSI	56, 89
data link protocol	30, 56, 71, 73, 104
data switch	**140**
data terminal equipment (DTE)	30, **33**
dBASE II	80
DEC	60, 62, 68, 70, 86, 118
dial-up links	24, 33, 36
digital research	76
Director	123
distributed computing	23
EBCDIC	68, 69, 70, 90
echo	151, 152
Edinburgh Regional Computing Centre	**82**
editor	21, 73
EIA	32
electronic mail	22, 105
electronic office	**22**
empty slot	117
encryption	57, 74
error correction	22, 71
error detection	22, 71, 92
error rate	100, 104, 130
Ethernet	116, 118
file sharing	103, 106
file transfer	23, 57, 60, **67,** 107
files	23, **67**
Filetab	80
fixed slot (*see* time division multiplexing)	

floppy disk 19
flow control 71, 96, 146
front-end processor 115
full duplex 38, 47

Gandalf 140
gateways 23, 99, 104, 125, 143

half duplex 38
HDLC 56, 147
host computer 24, 61, 71, 83, 84, 134, 140
hunt group 125

IBM 20, 53, 68, 70, 79, 86, 131
ICL **63**
IEEE 118
Informatics General 80
input/output (I/O) device 60, 75
Intel 19, 118
interchange circuit **32,** 49
interrupt 71
interworking 30, 56
IRMA 131
ISDN 130
ISO 24, 26, 36, 53, 68, 118, 148, 149, 154

joint network team (JNT) 82, 134

Kermit **87**
keyboard 149

large-scale integration (LSI) 14
leased line 24
local area network (LAN) 24, **97**
loop 111

macro 73

INDEX

mailbox	23, 144
mainframe	23, 30, 38, 63
Manchester Encoding	113
Microcom	87
microcomputer	11, **13**, 60, **63**, 153
micronet	24
microprocessor	15, 20
Microsystems Centre	77
minicomputer	13, 82
modem	30, **36**
morse code	27
MOS technology	19
Motorola	19
multi-drop	115
multi-tasking	21
native file transfer	87
NCC	60
network layer, OSI	56
networks	**97**, 129
– access methods	98, **114**
– broadcasting	104
– bus	98, 112, 115
– contention	102, 115
– deterministic	102, 117
– error rate	100, 104, 130
– gateways and bridges	23, 99, 104, 125, 143
– local area	24, **97**
– loop	111
– management	123
– performance	102
– resource sharing	22, 103, 107
– ring	98, 111, 117, **120**
– security	103
– speed	101
– star	110
– terminal	107
– topology	**110**
– transmission mode	112
– transparency	104, 107, 135, 147

– wide area 97, **129**
null modem 33, **45**

operating system 14, **20,** 70
optical fibre 112
OSI
  (see reference model for
  open systems interconnection)
Owl Micro-Communications 63, 144

PABX 130
packet switching 98, **130**
PAD 85, **132**
parallel transmission 27
parity 28
peripheral interchange program (PIP) 75
physical layer, OSI 56, 88
Planet **120**
polled network 38, 63, 71, 115
presentation layer, OSI 57
Prestel 144
printer 136
process control 115
programming language 21
proprietary network architecture **131**
protocol 24, 55, **87,** 129, 151
protocol converter 96
PSS 97, 132, 143
PSTN 24, 33, 143
PTT 129, 130, 144
public data networks 24, 132

queuing 125

Racal-Milgo 120
random access memory (RAM) 14, 17
read only memory (ROM) 17, 62
reference model for open systems
  interconnection (OSI) 24, 53, 87, 118, 142, 154

INDEX

register insertion	117
repeater	99, 120
reservation techniques	116
resource sharing	22, 103, 107
response time	64, 151
ring network	98, 111, 117, **120**
RS 232	30, 33, 49
scrolled display	60
security	23, 24, 103
serial interface	**28, 45,** 133
server	103
session layer, OSI	56, 89
signalling	35, 113
simplex	38
SNA	97, 131
speed matching	100, 102
spreadsheet	15, 21, 23, 60, 95
standards	24, 32, 51, 86, 149
star network	110
statistical multiplexer	139
Superbrain	86
switching	
– circuit	130, **140**
– packet	98, **130**
synchronisation	28, 30, 56, 113
synchronous transmission	**30,** 63, 71, 145
T-link	87
Tektronix	62, 85
telephone circuit	24, 36, 70, 102, 129, 144
Teletex	23, 98, 143
teletype	60
Telex	23, 142
terminal cluster	
(*see* terminal concentrator)	
terminal concentrator	63, 107, 115
terminal emulation	**59,** 78, 150
time division multiplexing	117

token passing   117
transceiver   118
transmission media   98, 99, **108**
transparency   104, 107, 135, 147
transport layer, OSI   56

Ultracom   63
Ungermann-Bass   118
utilities   **20,** 72, **75**

V-series recommendations
 – V.10   51
 – V.11   51
 – V.24   30, **32, 45,** 102
 – V.28   32, **35, 49**
value added network (VAN)   24
Videotex   23
virtual circuit   123
virtual file transfer   87

wide area network (WAN)   97, **129**
Winchester disk   19
window   72, 148
word processing   60
workstation   23, 105, 106

X-ON/X-OFF   85, 96, 147
X-series recommendations   132
 – X.3   132
 – X.25   84, 132
 – X.26   51
 – X.27   51
 – X.28   133
 – X.29   133
Xerox   98, 118
Xmodem   87

Zilog   19